Virtual Victorians

Networks, Connections, Technologies

Edited by
Veronica Alfano and Andrew Stauffer

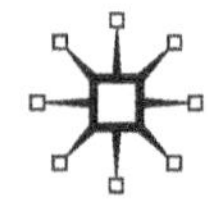

VIRTUAL VICTORIANS

Softcover reprint of the hardcover 1st edition 2015 978-1-137-39820-8

First published in 2015 by PALGRAVE MACMILLAN® in the United States—a division of St. Martin's Press LLC, 175 Fifth Avenue, New York, NY 10010.

Where this book is distributed in the UK, Europe and the rest of the world, this is by Palgrave Macmillan, a division of Macmillan Publishers Limited, registered in England, company number 785998, of Houndmills, Basingstoke, Hampshire RG21 6XS.

Palgrave Macmillan is the global academic imprint of the above companies and has companies and representatives throughout the world.

ISBN 978-1-349-48530-7 ISBN 978-1-137-39329-6 (eBook)
DOI 10.1057/9781137393296

Library of Congress Cataloging-in-Publication Data

Virtual Victorians : networks, connections, technologies / edited by Veronica Alfano and Andrew Stauffer.
pages cm
Includes bibliographical references and index.
1. Humanities—Electronic information resources. 2. Humanities—Computer network resources. 3. Humanities—Study and teaching (Higher)—Data processing. 4. Humanities—Research—Data processing. 5. Virtual reality in higher education. 6. Literature and technology. 7. English literature—19th century—Study and teaching (Higher) 8. Great Britain—Intellectual life—19th century—Study and teaching (Higher) 9. Great Britain—History—Victoria, 1837–1901—Study and teaching (Higher) I. Alfano, Veronica, 1983– editor. II. Stauffer, Andrew M., 1968– editor.
AZ195.V58 2015
025.06'0013—dc23

2014046289

A catalogue record of the book is available from the British Library.

Design by Amnet.

First edition: May 2015

10 9 8 7 6 5 4 3 2 1

Contents

List of Figures and Tables

Acknowledgments

Virtual Victorians was the brainchild of Meagan Timney and would never have come into being without her original vision. We owe her a great debt of gratitude.

Warm thanks also go to our anonymous reviewers, to our editors at Palgrave, and to all our contributors.

Introduction

Virtual Victorians

Andrew Stauffer

The title of this volume of essays, each of which attends to issues of representation and media change in nineteenth-century literary culture, evokes a series of gaps: between the real Victorians and the virtual ones, between sheer presences and imagined avatars, between the made-analog nineteenth century and the born-digital now. Indeed, such divisions increasingly animate our conversations about the Victorian era, as it fades decidedly into the past—now two centuries removed, all of its residents dead, its creaky media and communication technologies more distant and more strange each year. We might well ask, what now do the Victorians have to do with reality? Yet such alienation also brings an enabling perspective and a set of tools for research and thinking that might alter our vision of the nineteenth century. As time and technology make plain that our Victorian period will henceforth always be a simulation or constructed model, we can investigate more assiduously those networks of remediation. Put another way, we can exploit the virtual to make the past operational. At the same time, we are able to see more clearly the era's own immersion in virtuality, both optical and textual, as a result of its own novel technologies and networks.

"I was never out of England—it's as if I saw it all." In this line, the speaker of Robert Browning's "A Toccata of Galuppi's" catches the "as if" spirit of the virtual, its grounding in a multimedia evocation of the past. Hearing Galuppi's music, he sees eighteenth-century Venice rise up before him—the "Balls and masks," "mirth and folly," the

"Dear dead women, with such hair"—which exist for him now only as vestigial Victorian imaginings. Yet they rise up with a certain power before him, like so many of Browning's men and women, those virtual creations of the dramatic monologue form. "Mesmerism" is perhaps his most direct exploration of the relationship between virtual and real presences. Concentrating on the idea of his beloved at a distance, he seems through sheer effort of will "to have and hold / In the vacancy / Twixt the wall and me, // . . . Her, from head to foot," until, as he says, "I imprint her fast / On the void at last / As the sun does whom he will / By the calotypist's skill." Making reference to Henry Fox Talbot's recent photographic invention, the calotype, Browning suggests the turbid imaginative commerce between technology and presence for the Victorians, veterans of unprecedented media change. We can perceive the Victorian emphasis on virtual beings in its many post-Romantic hauntings, from Charles Lamb's dream children to Thomas Hardy's phantom horsewoman—all those revenants called forth, like spirit photographs, at the intersection of vision and mediation.

Our own virtual Victorians come to us out of the archives, which are increasingly available to computational analysis via digital surrogates and models of various kinds. The nineteenth century, and the Victorian era in particular, have been exceptionally rich target areas for digitization, given the plenitude of material produced by the industrial press, its freedom from copyright restrictions, its often visually rich texture, and its informational complexity as part of a rapidly changing media landscape. Early scholar-built digital archives, such as the Rossetti Archive, set the standard for online editions and led to the digital federation NINES (Networked Infrastructure for Nineteenth-Century Electronic Scholarship), which has peer-reviewed and aggregated work in this area. At the same time, commercial providers, such as ProQuest and Gale/Cengage, have been assiduously building digital collections of Victorian periodicals and newspapers, and the Google Books project has been amassing great ranges of Victorian digital content from academic library shelves. Libraries themselves are contributing growing amounts of digitized content (often focused on manuscripts and other unique collections), and new partnerships between digital humanists, librarians, and scholars have produced projects like The Shelley-Godwin Archive and Leigh Hunt Online. It is unquestionably an exciting time to be a Victorianist, with all of this material opening up before us in new ways, made available to new kinds of analysis.

As a result of these many digitization initiatives, Victorianists find themselves working in an increasingly rich and varied landscape of material, even as they are often contributors to the remediated archive of their period. Our first group of essays, "Navigating Networks," takes up some of the issues and opportunities of online search and the digitization efforts upon which it depends. We begin with Catherine Robson's essay, "How We Search Now," a report from the archives, both print and digital, where she has traced "the overt and covert surfacings" of Charles Wolfe's once-famous poem, "The Burial of Sir John Moore after Corunna" (1817). While giving us a master class in literary research methods, Robson reflects upon the "hybrid nature" of scholarly investigation today, which must rely on digital search as well as close reading of nondigitized materials. As expected, algorithmic phrase-matching across the large data sets of digitized nineteenth-century materials reveals unexpected appearances and citations of the poem. However, other evidence emerges from Robson's attentive research and reading in analog materials—in this case, memoirs of army chaplains from the Boer War to World War I. As a searcher, she finds direct citations; as a reader, she finds encrypted echoes and resonances in scenes of battlefield burial. Making an argument for a more theoretical consideration of what searching means across the partially digitized nineteenth century, Robson demonstrates that we all must be "double agents," able to mine large digital resources while also attending closely to textures of language only perceptible to a human reader.

Like Robson's essay, Ryan Cordell's contribution, "Viral Textuality," explores strategies of tracking the spread of particular nineteenth-century texts through networks of Victorian print. Cordell is interested in the mechanisms of circulation and reception that shape the life of a text: "its vectors of transmission, its diverse audiences, its many and varied modes of expression" as it appears across nineteenth-century American newspapers. Using Charles MacKay's oft-reprinted poem "The Inquiry" as one example, Cordell explicates its varied paths of transmission and transformation, which tell us much about the social and technological mechanisms of "going viral" in the Victorian era. He offers the metaphor of virality as a way of thinking about textual exchange and "rhetorical velocity," and gives a preview of the Viral Texts project, which will use reprint-detection algorithms to track specific reprintings of various texts and see more clearly the platforms and gatekeepers of the nineteenth-century press. At the same time, Cordell's essay suggests deep continuities between the structures of Victorian media and current social media practices. Not only do we need digital tools to study the past, he implies, but we also need to study past practices to illuminate our current digital environment.

In her essay on the Victorian writer Eliza Meteyard ("Silverpen"), Susan Brown shares with Cordell and Robson a primary interest in how we navigate the changing literary research environment. Brown gives us a metareport on tracking Meteyard via Google and the Orlando Project, seeing in their different data and perspectives some lessons for the construction of feminist literary history. How do we recreate the complex webs of text and context in which writers like Meteyard operated? Brown considers semantic tagging and linked data as paths toward new literary histories, yet she urges care and awareness of the various pitfalls of algorithmically displayed information. Her essay is in fact a call for greater scholarly participation in the building of scholarly corpora and tools, which too often reflect cultural bias or technologically induced assumptions that reinforce old ways of thinking. She urges us "to build inflection, orientation, flexibility, and difference into the emergent structures of the semantic web," allowing for the emergence of new representations of the past.

Alison Booth focuses similarly on the ability of new digital tools to reveal "documentary social networks" of the Victorian era. Using Frances Trollope as a case study, and with reference to her own Collective Biographies of Women (CBW) project, Booth shows how exposing representational networks to digital analysis can reveal unexpected patterns and "more nuanced interpretations of narrative structure." Reading Trollope's life across multiple biographical contexts (and with the aid of an XML schema, or taxonomy, of life events), the user of CBW perceives issues of representations, gender, and genre that would otherwise be occluded. Booth's project suggests the value of "midrange" reading, a kind of structured, lateral movement across similar literary works that combines embedded close reading (present in the markup language) and quantitative analysis. As with Susan Brown's essay, Booth's contribution argues for the interpretive payoff of scholar-built digital resources that move beyond the flat text of commercial databases and Google Books. Literary historians need to create the virtual Victorians they require.

Michael E. Sinatra's essay on representing Leigh Hunt's *Autobiography* also comes out of a digital project meant to represent a Victorian textual field in digital space. Sinatra reports on his extensive research into the evolution of Leigh Hunt's self-representations and consequent reputation through the early and middle decades of the nineteenth century. Hunt's position as a crossover figure between the Romantic and Victorian eras makes him an ideal figure for tracking a cultural transition visible in tight focus via his revisions of his own earlier work. Sinatra outlines his plan to expose Hunt's textual remains to scholarly study in a digital environment, a planned Leigh Hunt Archive that will provide access to these materials not only for reading, but also for data mining and visualization. This highly visual and collaborative archive will, says Sinatra, reimagine the digital scholarly edition as a place of experimental representations and speculative engagements. In this way, it takes its place alongside the writings and revisions of its subject author: Leigh Hunt was a master of virtual selves if there ever were one. Sinatra concludes with some meditations on the state of the digital humanities as an emerging field and the consequences of its emergence for the disciplines of editing and literary studies.

In her essay "Visualizing the Cultural Field of Victorian Poetry," Natalie Houston engages in a mode of digital reading that draws on computational data to reveal new patterns of publication. Focusing on library catalog records and anthology tables of contents, Houston offers an extended reflection on "data" in the humanities and then puts into practice her case for the value of quantitative visualization. Processing the metadata of Victorian-era poetry books in WorldCat, and thus turning lists into relational forms, Houston uses network analysis methods to reveal relationships between authors and publishers, organized by temporal and geospatial coordinates. She does a similar operation on tables of contents for our modern anthologies of Victorian literature, creating graphs and visualizations of the selections and their ordering. In her analysis of "the collective virtual Victorian reality created and used by scholars, archivists, and curators," Houston defamiliarizes Victorian poetry by making operational the metadata of its presentation, in its own time and in ours.

The second half of this collection, "Virtual Imaginings," gathers essays dealing with Victorian technologies of virtual experience. Here our authors are implicitly or explicitly outlining a prehistory of digital virtuality by way of explorations of specific Victorian cultural forms and their legacies. All of these essays take up some aspect of Victorian virtuality as such, with an emphasis on nineteenth-century technologies and the imaginative structures to which they gave rise. That is, while the first half of this collection emphasizes the consequences and opportunities of the digitized Victorians, the second half looks to media change from the spectacular "Panorama of London" of the late 1820s to early cinema around the turn of the century.

In her opening essay on "the virtual world of print," Alison Chapman approaches Victorian poetry from its position within periodicals, locating cultural resonances explicitly within the serial form. Like Brown with Eliza Meteyard, and like Booth with Frances Trollope,

Chapman uses a single, relatively neglected author as a heuristic, reading closely the poetry of Mary C. Gillington to explore "poetry's engagement with the conception of time embodied in periodical print culture, the hypertemporality of a virtual world that is date-stamped and yet cyclical." Chapman's essay establishes the nuanced scene of virtuality at work in serial publications, with particular attention to *Woman's World* under the editorship of Oscar Wilde. Explicating the layered periodical contexts of Gillington's pastoral cycle of poems, Chapman shows the ways "in which Victorian poetry engages with an immersive world, and the methods by which poetry both disciplines and disrupts the hypertemporality of serial print."

Peter Otto's contribution, "Artificial Environments, Virtual Realities, and the Cultivation of Propensity in the London Colosseum," explores the dynamics of immersion and participation in the virtual space created by the nineteenth-century panorama. Against a narrative of surrender to the reality effect of these technological spectacles, Otto shows that panoramas frequently engendered meditations on "scopic regimes," virtuality, and the relationship of technology to realism—all conducted within resolutely social spaces that were part of the overall effect. His case study is Thomas Hornor's Colosseum (with its "Panorama of London") "and the virtual realities that it constellated." For Otto, immersants in Horner's panoramic scenes "become cocreators—who through conversation, imitation, and display help shape the virtual worlds" in ways that prefigure computer-driven virtual spaces, social media economies, and affective environments of the present day.

Ruth Brimacombe is also interested in the virtuality effect created by Victorian pictorial technology: in her case, pictorial reportage printed as wood engravings in the periodical press of the later century. Brimacombe shows how Prince Albert Edward's tour of India in 1875–76 was extensively represented in images in the press that self-consciously attempted to create a virtual experience for the viewer. Linked explicitly to the modes and effects of the panorama, pictorial reports have all the hallmarks of the virtual experience: "the sensation of an immersive experience, the illusion of mobility, and the development of affective qualities." By manipulating perspective and scope, illustrators could actually produce more sophisticated virtual effects than photographers could manage, and yet their on-the-spot sketches maintained a link to reality that informed the scene of reception back home. As Brimacombe argues, "the contribution of illustrated journalism to the visual history of the nineteenth century needs to be fundamentally reassessed" in light of such cultural prominence. What's more, digital technology makes this newly possible by allowing for the easy integration of images into the scholarly conversation. Our virtual environments give us better purchase on the ones the Victorians themselves created.

Lisa Hager's essay on Scott Westerfeld's *Leviathan* series redirects our attention to what Simon Joyce has called "the Victorians in the rear view mirror," by way of young adult steampunk fiction and its refraction of Victorian technology. For Hager, "gender transgressions . . . and the technologies that make those transgressions possible" are preeminent in this trilogy of novels, centered as they are on the creation of a virtual Victorian world. Unpacking the shifting identities of the protagonists, Hager demonstrates the value of contemporary steampunk fiction in complicating the binaries often associated with Victorian gender roles. Aesthetics and technology—corsets and cogs—become the indices of the "steampunked" Victorian era, a virtual world in which new configurations of character and sexuality can be explored as heuristic devices for reencountering the boys' books and courtship novels of the period. Hager's essay reminds us of the continual reshaping of our Victorian imagination and the complexities of our constructions of the past.

The final essay in the volume, by Christopher Keep, turns to Rudyard Kipling's short story "Mrs. Bathurst" as a way of thinking about the virtuality effect of early cinema, a medium that encouraged awareness not only of the moving image but of the interface or frame in which it appeared. Kipling's story centers on one man's obsession with the filmed image of the title character, moving toward the camera and then out of the frame. Film's ability to suggest other places, other scales of time and presence, becomes a lodestar for memory and attention, and Keep connects this to Walter Benjamin's "optical unconscious" and the power of contingency in our experience of the virtual. As he argues, in a surprising reversal, "the virtual is defined not by its verisimilitude but by its apparent alterity," its emphasis on the intractability or withdrawal of that other world, a haunting reminder of the complex affective economies of our attraction to virtual Victorians.

A. E. Housman's Shropshire lad catches something of this spirit via "an air . . . / From yon far country," one that brings with it a vision of the past that is at first bewildering and ultimately wrenching:

> Into my heart an air that kills
> From yon far country blows:
> What are those blue remembered hills,
> What spires, what farms are those?

> That is the land of lost content,
> I see it shining plain,
> The happy highways where I went
> And cannot come again.

At the far end of Wordsworth's century, the Victorian poet austerely recasts the "sad perplexity" of his Romantic forebear by pitching the revisited landscape into the realm of the virtual. The hills, spires, and farms are "remembered" in the first stanza, but what shines plain is not the Wye valley or the Shropshire village, but "the land of lost content," with its "happy highways" now closed to foot traffic. Virtual paths are those upon which we cannot really travel: the past is the land of the lost.

And yet, as our tools for creating rich representations of Victorian literary culture expand in capacity, we have an opportunity to see that world shining plainer, bluer, maybe even happier than before. The virtual Victorian era that is emerging from our algorithmic searches, our digital editions and tools, and our data visualizations necessarily reflects that era's own navigation of changing media and information technology. What will we find in that "land of lost content"? The very term wobbles. "How long do you mean to be content?" asked Shelley's visionary double on the terrace at Pisa, soon before Shelley himself joined the world of virtual presences that would haunt the Victorian century so assiduously. We hope this collection of essays will suggest new ways of thinking about Victorian content—the archive and its affect, now and then.

Part I

Navigating Networks

Chapter 1

How We Search Now: New and Old Ways of Digging Up Wolfe's "Sir John Moore"

Catherine Robson

I begin not with the Victorian period but with our own millennium—with, in fact, the Millennium trilogy, the phenomenally successful series by Swedish crime writer Stieg Larsson, the first of which, *The Girl with the Dragon Tattoo*, was published posthumously in 2005.[1] Let me be more precise: I begin with the 2009 film adaptation of this novel, directed by Niels Arden Oblev, a movie that enjoyed widespread critical and commercial success and was later remade into an extremely faithful 2011 American version. Even if you have not read the books or seen the films, you have probably heard something about them, most likely something about their central figure, Lisbeth Salander, who has been hailed as both a postmodern heroine and a striking alternative icon for our times; certainly Noomi Rapace's performance in this role in Oblev's film creates an unforgettably powerful image. Salander is introduced to viewers as an unnervingly gifted isolate, a multiply-pierced and tattooed loner who is preternaturally skilled at penetrating the complex network of encrypted information systems in which she makes her living as a private investigator. In the movie's establishing phases, Salander's connection to her laptop is her most important relationship; her primary identity is that of a punk cyberhacker who exists most comfortably in webs of digitized data. To make this clear, Oblev presents numerous images of the top half of Rapace's face reflected in her computer screen. Lines of digital text run across her eyes and forehead, as if to illustrate that her visual and cognitive capacities are entirely merged with those of her Mac; the central processing units of brain and computer, we are to infer, work as one.

Later on we become all too aware of Salander's vulnerable corporeality. In a scene that viewers invariably find the most disturbing of the film's many violent sequences, she is the victim of a brutal anal rape by her state-appointed guardian. But instead of taking us to the dark interior of that abuser's bedroom, I move now to another badly lighted space, a location to which Salander is led by another older male figure in the film's denouement. On this occasion the man is an archivist and the place is an archive, the holding space of a large Swedish business conglomerate's records. To pursue her research into a series of murders and disappearances half a century earlier, Salander must walk through floor-to-ceiling stacks of cardboard files and handwritten ledgers, and lug box after box of documents to a little desk set amid the gloom. Poring over these papers, Salander starts to figure out the answer to the mystery but finds that there are more materials she needs to examine. When she returns to the chronologically ordered shelves to find the boxes for 1951, 1952, 1953, and 1954, Oblev repeats the visual strategy he employed earlier with Rapace's eyes and forehead. This time, however, he superimposes the green rectangular spines of the numbered files across the top half of her face. Back at the desk with its growing pile of faded receipts and reports, Salander opens her MacBook to cross-reference a photograph of the suspected murderer in an old company magazine with one stored on her hard drive. The verification is made, and the killer's identity is revealed.

This is the film's big "aha!" moment, but I do not draw attention to the scene because of any particular interest in the solution it presents. Rather, I am concerned here with method—with the ways in which the film takes pains to establish that the answer is found only because of Salander's ability to search and think in both new and old modes. Oblev underscores this point most obviously with those two composite visions of his heroine's face; via this pair of images, he shows that Salander's consciousness can merge with the protocols of digital *and* print media. Modernity may offer the skilled analyst incredibly sophisticated tools for gathering and interpreting information, but to make sense of the past's continued grip upon the present, a heroine for our times must also be able to engage with historical records in a more traditional fashion.

As it is with Salander, so it is with literary scholars today. This essay foregrounds the following fact: even with all our search engines and our Web proficiencies, our digitized media and our handheld electronic devices, sometimes we, like Salander, can only find out the truth—or what passes for truth in the world in which we operate—by looking closely at the box file, the handwritten ledger, and that assemblage of pieces of paper, the book. Most of us, I believe, thus spend much of our time operating as Salanderian double agents—that is to say, we conduct our research in both newfangled and oldfangled ways. These truths may seem self-evident, but I state them here so that I can proceed to the following question: what insights might be gained if we took conscious account of the hybrid nature of our current working methods? In our zeal to attain other, seemingly more urgent, goals, we may not comment on this hybridity in our published writings, but I suspect that many of us think about it a great deal and are aware of how it skews—or in some way affects—not only our findings but also what we say about them. The following pages represent an attempt to be candid about the use of different research methodologies and archives in my own scholarship, and to reflect upon some of the things I learned when I switched between them.

I should make it clear straightaway that I am not moving toward a simple valorization of the old at the expense of the new. This piece will not, for instance, feature a celebration of the romance or the erotics of the old-style archive, topoi that have found frequent representation in both academic studies and books and movies for wider audiences.[2] And neither shall I proceed in the other direction, which is to say, toward a wholehearted embrace of the actualities or potentialities of the digital humanities as currently practiced or imagined. While this essay supports the insistence on "bout[s] of methodological self-consciousness" that is both expressed and illustrated in one of the pamphlets produced by Stanford's Literary Laboratory, it makes no overtures to digital engagements of any complexity, such as those that make possible the modes of quantitative analysis practiced by Franco Moretti and his crew.[3] Indeed, rather than going "Beyond Search" (as part of the name of an earlier iteration of the Literary Lab has it), it stays with "search" for the simple reason that this is the command I assume the majority of us use most frequently with digitized materials. So, seductive though fantasies of revivification in the dust of special collections or the white heat of new technologies may be, my desires here are relatively restrained. In his review essay for NINES, "Digital Scholarly Resources for the Study of Victorian Literature and Culture," Andrew Stauffer urges us to consider "what relationship critical inquiry has to the archive of study."[4] I take this prompt as an invitation to ask what it means for our scholarship that our two distinct modes of searching link us to different archives, require different reading and analytical practices, and produce markedly different results.

The questions I am raising could be construed in various ways and worked out across very broad or extremely narrow issues. I give them specificity here by exploring them in the context of my own scholarly practices over the last decade or so, a period in which I was working on a book about nineteenth- and twentieth-century poetry memorization and pedagogical recitation. Much of this project, now published as *Heart Beats: Everyday Life and the Memorized Poem*, organizes itself around the reception histories of three short poems. To be perfectly frank, these portions relied to a considerable degree upon a stunningly banal research methodology: I spent a great deal of time putting lines or phrases from my chosen works into search engines and seeing where my hits would take me. This study, then, is fully a product of the "Age of Google." It would not have been possible for me as a single individual to have found quotations from such a wide and varied set of texts as I eventually cited without the benefit of digitized media and the means to search them. Given the exponential increase in digitization projects over the past ten years, it is also a book that would turn out rather differently if I started it today. At the time of writing this essay, I can see that many of my once hard-to-find, must-visit-the-specialized-library texts are now available at the touch of a few buttons. This strikes me both as an exciting development, when I consider it from the vantage point of future prospects for our discipline, and a rather depressing one, if I think about it in relation to the labors of my earlier self.

I was perhaps most conscious of the effects of my frequent oscillations between old and new modes of searching in 2004, when I was investigating the reception of Charles Wolfe's once widely memorized, but now largely forgotten, work of 1817, "The Burial of Sir John Moore after Corunna." I decided to make this poem about the rushed Spanish interment of a British hero of the Napoleonic Wars the subject of a case study because of an experience in the old British Library newspaper archive in Colindale in 1996. (This "outdated warehouse," to quote the BL website, was closed down at the end of 2013.) Back then I was curious about the backstory of Thomas Hardy's "Drummer Hodge," quite a different literary composition about a dead soldier from a later British war; I was combing through the columns of the *Dorset County Chronicle*, the *Dorset and Somerset*

Standard, and other southwestern periodicals of 1899 to see if they made reference to the death of a Dorchester drummer that Thomas Hardy mentions in a footnote to his poem.[5] I didn't find the drummer, but I did notice that the frequent brief obituaries and accounts of soldier casualties in South Africa in these newspapers often carried the same title: "No Useless Coffin." These three words, I then discovered, came from Wolfe's verses, and their repeated citation demonstrated that although this work was at that time new to me, it had once been intimately known by a broad community—so well known that it was indisputably the go-to poem for any description of a hasty battlefield burial. Because this current essay will pay close attention not just to those three words but also to many of the others in "The Burial of Sir John Moore after Corunna," I reproduce the poem here in full for reference:

Not a drum was heard, not a funeral note,
As his corse to the rampart we hurried;
Not a soldier discharged his farewell shot
O'er the grave where our hero we buried.

We buried him darkly at dead of night,
The sods with our bayonets turning,
By the struggling moonbeam's misty light
And the lanthorn dimly burning.

No useless coffin enclosed his breast,
Not in sheet or in shroud we wound him;
But he lay like a warrior taking his rest
With his martial cloak around him.

Few and short were the prayers we said,
And we spoke not a word of sorrow;
But we steadfastly gazed on the face that was dead,
And we bitterly thought of the morrow.

We thought, as we hollow'd his narrow bed
And smooth'd down his lonely pillow,
That the foe and the stranger would tread o'er his head,
And we far away on the billow!

Lightly they'll talk of the spirit that's gone,
And o'er his cold ashes upbraid him—
But little he'll reck, if they let him sleep on
In the grave where a Briton has laid him.

But half of our heavy task was done
  When the clock struck the hour for retiring;
And we heard the distant and random gun
  That the foe was sullenly firing.

Slowly and sadly we laid him down,
  From the field of his fame fresh and gory;
We carved not a line, and we raised not a stone,
But we left him alone with his glory.[6]

In 2004, when I started putting phrases from this poem into search engines, I was hoping to find out more about the poem's currency among British soldiers on active service in the nineteenth and early twentieth centuries—most notably in the two Anglo-Boer Wars (the First, 1880–81; the Second, 1899–1902) and the First World War (1914–18), but perhaps in the Crimean War (1853–56) as well. What I got instead was hit after hit after hit from texts related to a completely different conflict on the opposite side of the Atlantic. "The Burial of Sir John Moore," I thereby discovered, had been an important point of reference for individuals who were caught up in the horrible carnage of the American Civil War (1861–65). The chapter on this poem in *Heart Beats* examines six American citations of Wolfe's lines in considerable detail, so these cases will receive only cursory comments here—but for illustrative purposes, I'll give a brief summary of some of the types of materials I found.

Patrick Henry Taylor, a sergeant in the First Minnesota regiment, quoted (or to be precise, quoted with personal alterations) the "No useless coffin" stanza in an entry he made in his notebook after the Battle of Gettysburg; these three words also appear, unattributed, in a sentence written a year earlier in a journal kept by Kate Cumming, a nurse for the Confederate forces in Corinth, Tennessee, and within a dispatch printed in the London *Times* in 1863 and composed by an unnamed Canadian journalist from a hut hospital in Fredericksburg, Virginia. Captain Loren Webb of the Ninth Regiment of Illinois Infantry Volunteers included a parody of the poem's first verse in a letter he submitted as a "war correspondent" to an Illinois newspaper in 1862; in addition, phrases from Wolfe's third stanza, again with personal alterations, crop up in 1863 in the middle of a letter from William Rhadamanthus Montgomery, a young "Georgian sharpshooter," written to his "Aunt Frank" in Marietta from the front lines in Knoxville, Tennessee ("'No useless coffin encloses his breast, but he lays like a brave warrior taking his rest with only his blanket around him'").[7]

Although I was already aware at this time that the lost practice of mandatory pedagogical recitation excites far more nostalgic reverence in the United States than it does in Great Britain, I had not seriously considered broadening my survey of memorized poetry from a one-country to a two-country study until I stared this evidence full in the face. It revealed not just how widely Wolfe's verses had been disseminated across the length and breadth of the then-settled United States, but also something considerably more profound and unusual about the special properties of the memorized poem: these scraps of text showed, in moving detail, how individuals in pain summoned up and altered the lines of a work they had carried within themselves since school days at a time when they needed its consolatory power to transform their own particular desperate circumstances. My desire to include these citations and explain their contexts and significance thus sent me on an arduous multiyear journey to double the national coverage of my project—a journey that required me, among other tasks, to become knowledgeable about both the general history of American education and the specific history of its classroom recitational practices.

This happenstance, then, significantly changed the parameters of my study, but it also made me think concretely about the issue that concerns me here, the ways in which our findings depend on the mode of approach. The fact that I uncovered so many American wartime references to my poem is not surprising: US Civil War buffs, amateur and professional alike, were quick to embrace the opportunities afforded by the Internet, and seemingly every piece of paper connected to the conflict was put up on the Web at a relatively early point. Knowing this, I did not conclude that "The Burial of Sir John Moore" was more important in American theaters of war than in British ones. Instead, I faced the sobering truth that I would have to find some way to trawl the as-yet-undigitized archives of Britain's nineteenth-century military history that would not swallow the next thirty years of my life. And so I then set about my research in the old-fangled way. I wrote to the archivist at the National Army Museum in London asking whether he had ever come across any references to Wolfe's poem in the personal documents of Crimean War combatants. After a little while, he wrote back to say no. I was not entirely convinced, but I decided, on reflection, to let it lie; after all, although "The Burial of Sir John Moore" swiftly became a general favorite from the 1820s onward, the pedagogical systems that would make it a British classroom standard later in the century were simply not in place during the childhoods of the soldiers who fought at Alma, Balaclava, Inkerman, and the rest.

Yet I knew that this was not the case for those volunteers who traveled to fight the Boers toward the close of the century. The working-class men who constituted the majority of the troops in these two wars comprised, for the first time in the history of the British armed forces, a literate soldiery who had received six to seven years of schooling that, for most of them, had included some compulsory memorization of verses. The poetic dimensions of the second South African conflict had already been the subject of a literary historian's attention: Malvern Van Wyk Smith's study *Drummer Hodge: The Poetry of the Anglo-Boer War 1899–1902* appeared in 1978. This excellent survey confirms that many of those who composed verses during the war were at least casually acquainted with Wolfe's ode:

> Much of the poetry . . . written by or about men at the front falls into the cadences of the popular recitational ballads and music hall tunes of the time. Indeed, a glance through a few of the well-known school-boy anthologies of the period, Henley's *Lyra Heroica* or Langbridge's *Ballads of the Brave*, will reveal the models of most of the Boer War verse produced both at home and at the front. Chief of these was the iambic quatrain of the border ballad . . . Developed during the nineteenth century into a veritable cult by the infectious efforts of writers such as Mrs. Hemans ("The boy stood on the burning deck"), the simple narrative ballad was the most readily available mode for Tommy poets . . . Another, and very popular, variation was to substitute an anapestic rhythm for the genre's basically iambic metre, as in Wolfe's "The Burial of Sir John Moore." (158–59)

So the poem's formal structure had been a reference point for soldier-poets and others. But was there any evidence that its content had played the same kind of role for individuals in South Africa as it had in the American Civil War? Had the words and ideas of the poem similarly offered solace for those involved in battlefield burials of fallen friends and comrades, or at least provided them with some sort of framework within which to express their feelings about the experience? *Drummer Hodge* makes no mention of such usages, so I wrote to ask Van Wyk Smith whether his research had uncovered any citations that he had elected not to include in his text. He did not remember; his boxes were in storage; he would look when he could. I waited until I heard further word, but when it came, it too was negative. What next?

At this point I did what we do when a search founders; I decided to alter its parameters. What if I moved my attention to those who had a guaranteed relation with versions of the scene depicted in Wolfe's poem? Members of the chaplaincy are and were the only army personnel directly charged with the responsibility of officiating over the funerals of casualties; their writings, it seemed to me, offered the best opportunity of finding my quarry. And the dates fitted: from histories of this branch of the military, such as Brigadier The Rt. Hon. Sir John Smyth's *In This Sign Conquer: The Story of the Army Chaplains*, I learned that it was not until the Boer Wars that there were sufficient numbers of clerics in the forces to make it at all probable that an ordinary soldier would receive a "proper" burial. Wellington had complained in 1811 that he had only one active chaplain within his entire army; at the outbreak of the Crimean War, the roster still stood at a single clergyman for 26,000 men. Evangelical enthusiasm, stirred especially by William Howard Russell's frontline dispatches for the London *Times*, worked quickly to ameliorate the situation during this conflict; other moves, at subsequent points during Victoria's reign, broadened the number of religious denominations represented within a newly formed Army Chaplains' Department. By the end of the century, Roman Catholic, Presbyterian, Wesleyan, and Jewish soldiers had some chance of finding a man of their own faith within the predominantly Anglican chaplaincy. Always depending upon the exigencies of individual circumstances, it was thus increasingly likely that a uniformed clergyman would conduct a holy service over the interments of private soldiers as well as those of officers.

To be sure, I was aware that in shifting the focus to this specialized group, I was giving up the chance of finding poetic citations by individuals who had gone through the mass system of public elementary education with which my study was most concerned. While the socioeconomic backgrounds of chaplains varied to a certain degree (nonconformist ministers were generally less likely to hail from the "officer class," but this was not always the case), these bearers of the Word had usually attained higher levels of literacy than most of the rank and file. And there were other potentially problematic divergences too. The religious and professional connection of clerics to the men they buried was inevitably conditioned by feelings and sentiments other than those demonstrated in most of my texts from the American Civil War; chaplains may have worn khaki like ordinary soldiers, but they were not normally subject to the emotional ties that bound men in the ranks together—men who were often friends and also, on occasion, family members. But, I temporized, these writings would offer me something new: the opportunity to gauge the influence of Wolfe's poem in a different affective zone. What I actually found was arguably more interesting. I discovered that I could detect traces of "The

Burial of Sir John Moore after Corunna" in ways that a computer search engine could not.

The British Library catalog revealed that there were numerous Boer War clerics who had given literary form to their experiences. I ordered up every title that seemed in the least promising and then sat back to examine the haul. Van Wyk Smith's book had already told me that men of the cloth safely ensconced in their dioceses back in the British Isles had been especially prone to compose piously belligerent verses at this time. The volumes I perused, however, were memoirs of those who had been more intimately involved with the conflict; bombarded by a mass of experiential and frequently distressing detail, these men found it more difficult, though not always impossible, to construct abstract formulations that glorified the fight and found meaning in the loss of human life. But did these chaplain memoirs include passages that brought individual corpses and burials into plain sight? Well, the answer was yes, but the relevant scenes represented a hermeneutic challenge for a critic on the hunt for the trace of a particular literary work. Unsurprisingly enough, interment scenes were usually infused by the spirit of redemptive theology; lines from the "Order for the Burial of the Dead," and from scripture more widely, tended to be at the front of chaplains' minds as they constructed these episodes in their own prose. And yet in certain instances I began to feel that I could sense the presence of another text as well, one that had more immediate relevance to the specific circumstances of a soldier's burial. Unlike the service from the Book of Common Prayer, or assemblages of biblical citations, this was a text that provided a consolatory script for those on military terrain, celebrant and mourners alike—men who were forced to inter their dead in makeshift conditions and had to walk away at the end of the ceremony, leaving the dead in an alien land.

As the remaining pages of this essay will reveal, my hours with the padres in the British Library were rewarded with a single direct citation from Wolfe's poem, albeit without attribution. Yet I also discovered that the poem inhabited chaplains' writings in a less tangible yet pervasive form. In making this assertion, I am aware that I run the risk of claiming that the poem influenced the work of writers who may have had little or no conscious awareness of its existence, or, more likely perhaps, who might respond that their representation of a battlefield burial necessarily runs parallel to "The Burial of Sir John Moore": the very fact that individuals found the ode a fitting resource when they laid a comrade to rest is itself grounded in the fact that the description of a military interment therein, for all its local color, comes close to archetypal. When all is said and done, Wolfe's poem faithfully records a sequenced event structured by ritual. But I am emboldened to make my argument nevertheless. While the ubiquitous presence of the poem in juvenile education and thus in British culture more widely during this period provides the general supporting context, I stake my claim on localized close readings. Perhaps one day our digitized search functions will be sufficiently sensitive to deliver the kind of findings I describe below, but for the moment, at least, the acts I performed to locate these passages seem to lie firmly within the province of the human brain.

Sometimes the appearance of certain words and concepts, or the characteristic form of individual clauses, indicated to me that the poem was close by; at other junctures, I credited its influence if I detected a particular intensification of literary language, a reaching toward Wolfe's poetic techniques and methods in what was otherwise a medium governed by generic prose forms, such as reportage or adventure narrative. Suddenly sentences began with adverbs; the incidence of alliteration, assonance, and consonance was abruptly and appreciably amplified; far more likely was I to find inversions of normal word order and reliance upon parallel constructions; far more stress was laid on the rhythm of clause and phrase. That such stylistic devices may simply have been conjured up by the solemnity of the topic was always a possibility; furthermore, the actual or implied presence within the scene of what one chaplain calls "the sublime language" of the burial service itself seemed to call upon its witnesses to imbue their own prose with as much dignity and gravity as they could muster.[8] And yet I still think I was right to hear the presence of Sir John Moore's missing obsequies even when overt quotations from the poem were absent.

The Reverend Robert McClelland's 1902 memoir, *Heroes and Gentlemen: An Army Chaplain's Experiences in South Africa*, provided the most interesting case in point, for it makes both direct and, I will claim, indirect reference to the ode. Predictably enough, given the title of McClelland's text, this Scottish Presbyterian is concerned throughout with establishing the gentility of British fighting men of all ranks; that he should turn to Wolfe's poem to emphasize that final respect is due to the private's corpse just as it is to the officer's repeats a move I had seen frequently performed within American Civil War writings. In his description of the first funeral he officiates in South Africa, some thirty-five pages into his memoir, McClelland closes the brief and unemotional account of the interment of men who have died of wounds and disease in an army camp with a marked (if slightly inaccurate) citation: the rites for soldiers of other denominations completed, "our turn came, and we stood by while the blanket-wrapped figure was lowered into its grave on the veldt; the Bible was read, prayer said, and 'we left him alone in his glory' [sic]" (35).

When the chaplain is next called upon to conduct a burial service, it is after the British defeat at Spion Kop and thus under far more traumatic conditions. This time "The Burial of Sir John Moore" seems to haunt McClelland's text as a structuring memory rather than an acknowledged presence. At the end of "A Battle and Its Graves," a chapter divided equally between accounts of the fighting and its aftermath, the chaplain describes what was for him the most affecting part of the experience—the appearance of the three Scotsmen he had to bury:

> But above all, I can never forget the faces of the dead, turned up as if in appeal to heaven, the lips slightly parted: the first shot through the temple, the second below the breastbone, and the third had a ghastly wound as if from an explosive bullet in the chest. We buried them side by side in their martial clothing as the sun was setting, reading the twenty-third psalm and engaging in prayer. It was touching to see their belongings turned out, some recently sent from home, and some curios for the loved ones when the war was o'er; alas! We carved a cross on the fence, and raised a cairn, and there they rest 'till the resurrection morn. (151)

Is it McClelland's demonstrated knowledge of Wolfe's lines that prompts him to elevate the soldiers' rumpled and, at least in one of their cases, horribly bloody, khaki uniforms into "martial clothing"? If so, the memory of the poem then subsides for a spell as the individual identities of the three dead men assert themselves. Their interior lives, their pasts, and their now-unachievable futures with family and friends back home are represented in the "touching[ly]" distinct contents of their pockets and packs. But if actuality is at war with poetry in this account, then poetic stylization wins the battle even before this sentence closes: "o'er" (a word Wolfe uses three times) and "alas!" signal McClelland's generic shift. Yet in his last statement we see the chaplain turning to poetry for something other than the consolation or distancing that ennoblement can bring. The penultimate line of "The Burial of Sir John Moore," with its regret that "we carved not a line, and we raised not a stone," seems to suggest a certain course of compensatory action to McClelland. Although in this passage's final clause the clergyman returns to the bedrock of his Christian faith and its promise of resurrection, his move to commemorate is expressed, in retrospect at least, as an improvement upon the lack recorded in Wolfe's poem. How else to read "We carved a cross on the fence, and raised a cairn"?

McClelland's memoir, then, allowed me to analyze both overt and covert surfacings of "The Burial of Sir John Moore." Today, if I made allowance for the slight citational error, I would be able to find its instance of direct quotation in very little time at all: the Bodleian Library's copy of this text has been available through Google Books since January 10, 2008, albeit in snippet form only. And no doubt, in this hypothetical scenario, that digitally mediated engagement with the text would then encourage me to track down a physical edition of the book, and, reverting to my oldfangled techniques of detection, I would subsequently come across the second passage I have just quoted and perform the kind of reading I provide above. This, of course, is our default methodology—not an either/or, but a continual switching between; we use both methods in tandem to support or advance the developing search as the need arises and are thus all Salanderians now.

So this essay's point is largely made, but let me hammer it home a little further with a few more examples of subterranean Wolfian traces in the as-yet-undigitized writings of arguably the most prolifically literary clergyman in uniform to date, the Wesleyan Owen Spencer Watkins. *Chaplains at the Front: Incidents in the Life of a Chaplain during the Boer War 1899–1900* (1901) was the title that first attracted my attention in the BL catalogue, but I was soon immersed in his 1899 volume, *With Kitchener's Army: Being a Chaplain's Experience with the Nile Expedition*, and *With French in France and Flanders: Being the Experience of a Chaplain attached to a Field Ambulance* (1915) too; Watkins also wrote a history of Methodist ministry in the army and a couple of novels inspired by his time overseas with the forces. Although it is not possible in this chaplain's case to establish definitive knowledge of "The Burial of Sir John Moore," I would argue that all three volumes about his tours of duty contain echoes of the poem; the countless and often interchangeable burial scenes therein are patterned after a familiar form. Following the transmutations of these references in the course of fifteen years of service, however, offers an unusual opportunity to see the increasing stress that is placed on a specific poetic archetype over time; ultimately, as we shall see, the ode's consolatory power is destroyed by the incomparably horrible and unprecedented circumstances of modern warfare.

Wolfe's poem seems particularly present in the author's mind when in his first book he describes the funeral honors performed in 1898 for one of the most celebrated commanders of the Victorian era, General

Gordon, who had been killed some fourteen years earlier. "Silently we tied up to the quay," begins Watkins, before proceeding to enumerate the various marks of respect then paid to Gordon's memory (224). Despite the fact that these include a twenty-one-gun salute, the playing of the "Dead March" from Handel's *Saul*, a toll for the brave, a psalm of praise and thanksgiving, a few words of scripture, a short memorial prayer, some extracts from the burial service, and the Lord's Prayer, Watkins seems determined to see this event as a repetition of another general's burial, some eighty-nine years earlier. "No elaborate service could have moved us," he writes, "like this plain act of worship on the Nile bank, close by where Gordon died" (229). It is perhaps in its account of the leaving of the hero, after "the Soudanese band played 'Abide with me,'" that the record becomes most evidently infused with poetry (229). After the Battle of Corunna, the British forces, in flight from the French, had good reason to depart expeditiously; here there is no such reason, but it seems important to Watkins that he create a sense of a rushed farewell:

> All too soon the bugle sounded, and we had to return to the boats, but we went with lighter hearts than we had come, for now our duty to the dead was done, and a stain on the British name wiped out for ever. We left him with the flags he had fought for flying over the place where he fell. (234)

In a positive welter of anapests and alliteration, Gordon is left alone with his glory, and the wrong of his ignominious earlier burial is made right—via a few calculated misrepresentations—through the encrypted presence of Wolfe's poem.

Descriptions of Boer War burial scenes in *Chaplains at the Front* also seem structurally and thematically haunted by the ode:

> In the darkness we buried them in the cemetery just outside the town; no light was permitted lest we should draw the enemy's fire upon us, so with only the stars to light us, the words of the service, which daily were becoming more and more mournfully familiar, were recited over the grave. (103)
>
> Sadly but proudly we laid our dead in their last resting-place in the shady dip below Waggon Hill . . . Then worn out and sad we returned to our camps to talk of the fine fellows we had known and served with so long, but whom we should meet no more upon earth. (131–32)

Stars, not "the struggling moonbeam's misty light," shine upon the first mournful scene; no "lantern dimly burning" can be allowed for this night burial, for "the enemy's fire," "the distant and random gun / That the foe was sullenly firing," poses too great a danger; "few and short" are the prayers "recited over the grave." In the second, "slowly and sadly," "sadly but proudly," Wolfe and Watkins lay their dead to rest and finally walk away.

Moving from Watkins's representation of his days in South Africa to his account of what he witnessed in the first months of the Great War is a disturbing and dislocating experience. To rank military conflicts according to their degree of awfulness is always to run the risk of displaying insensitivity to individual experience; as Van Wyk Smith remarks, "all wars must seem equally cruel to the people involved in them" (3). Given the Boer War's 1 percent casualty rate, it was statistically fairly unlikely for any given British soldier not to survive it. Yet this fact would have meant little, if anything (to follow the reasoning of Sissy Jupe in *Hard Times*), to the friends and families of the men who died. For Owen Spencer Watkins, however, the scale of death on the veldt meant that his time, and thus his representation thereof, was only infrequently taken up by the performance of burial services; there was both occasion for other pastoral duties and the opportunity to observe with interest an unfamiliar land. Back alongside men in combat fourteen years later, however, the seasoned chaplain found few such moments in Northern Europe.

The bulk of *With French in France and Flanders* concerns itself with the devastating events connected with the retreat from Mons. Beginning on August 24, 1914, this movement was rendered necessary when all four divisions of the British Expeditionary Forces were outflanked after their first twenty days at war; while there were around 5,000 German casualties, it is estimated that over 15,000 British soldiers were killed or wounded, or went missing, in the battle and its aftermath. Watkins begins his narrative with the rhetorical gesture of inexpressibility that Paul Fussell has identified as a signal characteristic of the literature of the Great War: "I cannot describe it," he states. "It will not bear thinking about" (17). And like many others, too, Watkins conjoins this confession of the impossibility of putting into words or even reviewing his experiences with his conviction of their essential indelibility: "it has left a mark on our hearts and memories which nothing can efface" (17). Variations of these statements recur at regular intervals throughout the text: "the whole can only be realized by one who has himself passed through a similar experience—I cannot describe it" (30–31); "I will not harrow your feelings by attempting to describe it—it was horrible beyond words" (35); "things too dreadful to be put in black and white . . . of these things I must not and cannot write" (88).

After Watkins's professions of his inability to convey the chaotic horror of thunderous explosions, shattered bodies, and the terror suffered by appallingly wounded men, the frequent burial scenes that appear in this narrative stand out as familiar and describable oases. Such an effect is an understandable result and probably a faithful representation of actual circumstances. Any burial is in itself usually an indication of a momentary respite from immediate danger, and Watkins and his compatriots are only able to dig a grave and conduct a service under conditions of relative calm. And yet even once this is acknowledged, it is clear that there is something intrinsically comforting to the chaplain about these episodes; not only can he repeat a crucial religious ritual with its verbal promise of resurrection, but he can also return to his habitual rhetorical strategies. The burial scene I quote below is one of four or five of comparable length and near-identical form and content:

> [My] sad task was . . . to lay to rest a brave lad of the Duke of Cornwall's light infantry. We buried him in the garden of the house into which the wounded had been collected; the grave had to be dug in the dark—to show a light would have cost the diggers their lives—the service was read by the light of a lantern shrouded with sacks; the volleys fired over his grave came from the enemy, for just as the service concluded the Germans made another attack on our trenches, and we had to retire hastily to the house, to avoid stray bullets which were coming our way. (116–17)

Precarious as these interments are (other services are disrupted by "shrapnel . . . bursting on the hill" [54] or "deafening . . . artillery fire" [83]), they offer Watkins in retrospect both religious and literary stability. Burying the lad "darkly at dead of night" by "the lantern dimly burning," and hearing the "discharge[ ]" of a "farewell shot" in the enemy's fire, allows him to shape his text according to a mournful, but knowable, poetics of war. Far better this than the mass burials he rushes past in his narrative ("in one grave we buried two officers and eighteen men, and altogether we buried forty-one" [65]), or, still worse, the brief notices of those whom they cannot stop to help at all ("these dead and other German dead that we passed by the roadside, we were obliged to leave for the peasantry to bury" [67]). Watkins is certainly not averse to abstraction in his prose—he is, for instance, quite capable of alluding to "the sorrowful fruit of the battlefield" (127)—and even his confessions of inexpressibility are apt to take on the most "literary" of stylizations, as in this quasi-Shakespearean gesture of disgust: "I cannot write of the horrors I saw . . . the smell of blood—faugh! I have it in my nostrils now as I write" (115, original ellipses). But what he encountered on the roads back to France, it seems, eventually took away either Watkins's desire, or his ability, or perhaps both, to make further record of a "Chaplain's Experiences" after the fact. "Of the fighting, who can tell—God hides it with the veil of His darkness, and in this we see His mercy," he states near the end of *With French in France and Flanders* (127). Although Watkins was in active service throughout the war (after Mons he was present at the battles of the Marne, Aisne, Ypres, and the Somme offensive, and then was promoted to assistant principal chaplain to the Third Army) and was mentioned in dispatches on four separate occasions, he wrote no more about what he had seen.

Until now, the story of the days I spent with reverend gentlemen in uniform in the summer of 2004 has been largely untold. My failure to notice in a timely fashion that I had written some 50,000 more words than were permitted in my book contract required me to perform the surgical removal of great swathes of the manuscript; one bleak week in 2009 saw McClelland, Watkins, and many others from various sections of the project hit the cutting-room floor. But one of the great consoling fantasies of scholarly labor, howsoever fangled, is that there is no such thing as total loss; it is always possible to tell ourselves that no road led to a dead end, that the journey taught us something useful even if its findings never found their way into print. I hope that the opportunity this essay has given me to revisit a number of archives, to remember what happened when I used "search" and when I could not, has not merely played a personally recuperative role. Like Lisbeth Salander, I believe it is important to discover where the bodies were buried; it is also worthwhile, I think, to document how we managed to find them.

Notes

1. I thank Ryan Fong for directing my attention to this film and suggesting I examine its archival scenes.
2. An archive discovery scene (of a handwritten draft of a letter in a book in the London Library) in A. S. Byatt's *Possession* has come to represent a novelistic locus classicus, but see also, among many other possible choices, Derrida, Steedman, and Mavor.
3. Ryan Heuser and Long Le-Khac, "A Quantitative Literary History of 2,958 Nineteenth-Century British Novels: The Semantic Cohort Method," *Pamphlets of the Stanford Literary Lab* 4 (Stanford: Stanford Literary Lab, 2012), 30.

4. Andrew Stauffer, "Digital Scholarly Resources for the Study of Victorian Literature and Culture," *Victorian Literature and Culture* 39 (2011): 294.
5. See my "Girls Underground, Boys Overseas: Representations of Dead Children in Nineteenth-Century British Literature," in *Dickens and the Children of Empire*, ed. Wendy Jacobson (London: Macmillan, 2000), 116–18.
6. I take the text of Wolfe's poem from the *Remains of the Late Rev. Charles Wolfe*, ed. the Rev. John A. Russell (Hartford, CT: Huntington, 1828), 29–31.
7. For further information on all these citations, see my *Heart Beats: Everyday Life and the Memorized Poem* (Princeton, NJ: Princeton University Press, 2012), 191–218.
8. Owen Spencer Watkins, *Chaplains at the Front: Incidents in the Life of a Chaplain during the Boer War 1899–1900* (London: Partridge, 1902), 47.

CHAPTER 2

VIRAL TEXTUALITY IN NINETEENTH-CENTURY US NEWSPAPER EXCHANGES

*Ryan Cordell**

N.B. Additional images associated with this chapter are housed in the digital annex at www.virtualvictorians.org.

INTRODUCTION

On February 8, 1862, the *Ashtabula Weekly Telegraph* (Ashtabula, Ohio) published on its back page a list of fourteen "Health Hints—Follies."[1] These follies, attributed to a "Dr. Hall," critique ideas about diet and exercise ("1. To thing [sic] that the more a man eats the fatter and stronger he will become"), education ("2. To believe that the more hours children study at school the faster they learn"), home organization ("5. To act on the presumption that the smallest room in the house is large enough to sleep in"), and even ethics ("7. To commit an act which is felt in itself to be prejudicial, hoping that somehow or other it may be done in your case with impunity"). This snippet exemplifies the listicle genre, which is often associated with popular content

* Thanks to my co-PIs David Smith and Elizabeth Maddock Dillon for their contributions, both technical and theoretical, to the Viral Texts project. Thanks also to project research assistant Kevin Smith, who helped me track down many of the references here from our very large database of nineteenth-century reprinting. Finally, thanks to project intern Laura Eckstein of Haverford College, who prepared the maps included here.

online in the early twenty-first century, but which was also common in nineteenth-century newspapers.[2] The "Follies" piece is, on its face, quite conventional and unassuming, but it was one of the most widely reprinted newspaper snippets of the nineteenth century, appearing at least 136 times in newspapers and magazines between 1862 and 1899.[3]

By the time "Follies" was reprinted in *The Appeal* (Saint Paul, Minnesota) on December 23, 1899, it had been substantially changed. The original list had been reduced from fourteen numbered to seven unnumbered items, while words and phrases from the items themselves had been altered. The article's title—which was to be read as the leading clause for each list item—had become the more direct and blunt "You Make a Mistake" (i.e., "You make a mistake . . . in eating without any appetite, or continuing to eat after it has been satisfied, merely to gratify the taste").[4] But the essence of the piece remained the same, having wended through newspapers and magazines across the country for decades, filling a few inches of column space in the *Caledonian* of St. Johnsbury, Vermont (April 25, 1862), the *Pacific Commercial Advertiser* of Honolulu, Hawaii (June 26, 1862), the *Big Blue Union* of Marysville, Kansas (August 15, 1863), the *Arizona Miner* of Fort Whipple, Arizona (February 29, 1868), the *Morning Star and Catholic Messenger* of New Orleans, Louisiana (May 8, 1870), and the

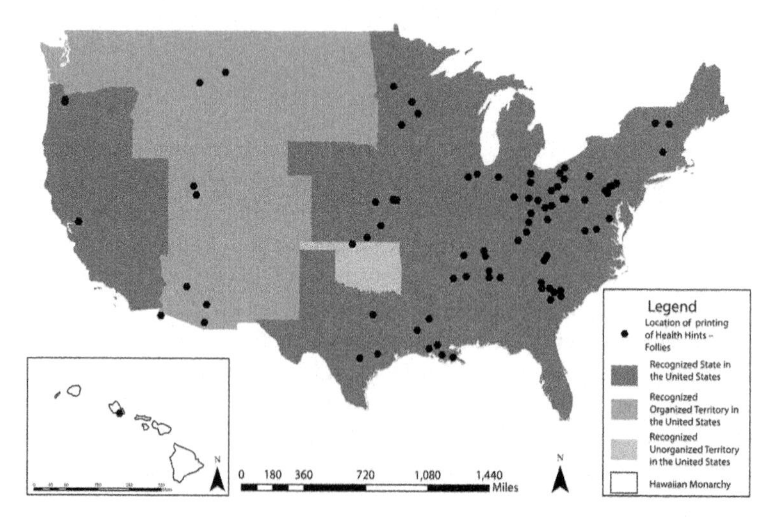

Figure 2.1 Reprintings of "Health Hints—Follies" in nineteenth-century American newspapers. The state boundaries here are from 1870, about midway through the text's life in the press, though these boundaries shifted considerably during the decades the text circulated. This map was prepared by Viral Texts project intern Laura Eckstein, an undergraduate at Haverford College.

New Ulm Weekly Review of New Ulm, Minnesota (April 23, 1879), to name just a few. In fact, this seemingly ephemeral snippet appeared in approximately 20 percent of the nineteenth-century newspapers now collected in the Library of Congress's Chronicling America archive. Considered in aggregate, as a widely successful newspaper selection, this piece is not simply a strange curiosity (as it might seem in the context of any individual newspaper) but is instead a telling example of popular newspaper literature. We might say "Health Hints—Follies" went viral, 150 years before social media.

These reprints of "Follies" were uncovered by the reprint-detection algorithm developed by my colleague, David A. Smith, for the Viral Texts project at Northeastern University. Our broadest aim for Viral Texts has been to "better understand what qualities—both textual and thematic—helped particular news stories, short fiction, and poetry 'go viral' in nineteenth-century newspapers and magazines."[5] We have written in detail about the development of this reprint-detection algorithm and about our experiments with different iterations in attempts to detect the most authentic reprints while omitting as many false positives as possible.[6] In this essay, however, I want to reflect on how the modern "viral" metaphor might contribute to theorizations of nineteenth-century reprinting practices. The word *viral* is necessarily anachronistic when applied to the nineteenth century; biological viruses were not themselves named until the century was nearly over. The "viral media" metaphor is even more recent, linking the spread of content or information to our understanding of contagion and containment. Did antebellum poems, fiction, news stories, travel accounts, trivia, and jokes "go viral" in any way analogous to the tidbits that tile our Facebook walls today? It is of course a stretch to call the antebellum newspaper "the Buzzfeed of its day"—but how much of a stretch, and is it a stretch worth making?

While it is a twenty-first-century neologism, I will argue in this essay that virality can provide a useful comparative frame for thinking about the exchange of texts in nineteenth-century newspapers and magazines. J. Gerald Kennedy notes briefly in his introduction to *Poe and the Remapping of Antebellum Print Culture* that "first-hand news reports from 'correspondents' circulated with viral alacrity, even though no organized system for gathering and distributing news yet existed," but he does not further develop an idea of how the viral metaphor might illuminate textual exchange within the "more leisurely and inconsistent pattern" of early nineteenth-century circulation. By applying recent theorizations of online virality to these historical mass media, however, we can identify common aspects of content sharing across historical periods, isolating cultural, aesthetic, and literary values that persist from print to Web. Perhaps most importantly, a viral theory of textuality foregrounds circulation and reception, describing not static textual objects but instead the ways texts moved through the social, political, literary, and technological networks that undergirded nineteenth-century print culture. Indeed, I would propose that a robust theory of virality, developed from corpus-level text analysis, offers a critical perspective that allows us to understand nineteenth-century circulation not simply as "arbitrary, irregular transmission of tales, essays, and poems," but instead as a complex but comprehensible system of textual exchange and influence.[7] Notions of virality highlight the ways texts create, sustain, or sometimes sever network connections.

Finally, virality offers a theoretical frame focused on textual similarity rather than originality or novelty—a necessary critical reorientation, I will argue, for the study of literature in mass media. The central fact of either nineteenth-century reprinting or twenty-first-century retweeting is repetition—the way ideas, instantiated in media artifacts, are inscribed and reinscribed in culture. In many ways, virality is another name under which to discuss bibliography. In essence, both virality and bibliography describe how often a given text was reprinted and in what venues. Unlike an enumerative bibliography, however (which focuses primarily on accounting for witnesses), or a descriptive bibliography (which attempts to account for the textual differences among witnesses), accounts of a text's virality would focus on its social life and rhetorical power. How far and in what forms did it spread? In which communities did it circulate? How was this text modified, remixed, responded to, or commented upon? To what extent did this text saturate a given network? How does the spread of this text compare with that of others? And, finally but perhaps most importantly, what textual, thematic, or stylistic features allowed this text to be easily shared? The lens of virality privileges questions of circulation, reception, and comparison over questions of textual authority; it elevates the social over the stemma. As an interpretive frame, virality also requires a more capacious understanding of "the text," including in its ambit both bibliographic witnesses and a penumbra of related, often fragmentary texts that speak to the social, literary, or historical impact of the original item.

DEFINING *VIRALITY*

To "go viral" today is to be widely and quickly shared—to move rapidly from obscurity to ubiquity—through interrelated platforms such as YouTube, Facebook, Instagram, Twitter, and Tumblr. In *Going Viral*, Karine Nahon and Jeff Hemsley offer this useful definition of *virality*, which they describe as:

> [A] social information flow process where many people simultaneously forward a specific information item, over a short period of time, within their social networks, and where the message spreads beyond their own [social] networks to different, often distant networks, resulting in a sharp acceleration of the number of people who are exposed to the message.[8]

In short, to call something "viral" is to note the speed, scale, and sociality of its spread. The pop star Psy, while not obscure in Korea before the appearance of his song and music video "Gangnam Style," became a global sensation in a matter of weeks after that video was viewed millions of times on YouTube.[9] We could list many more examples, from the amateur "Yosemite Mountain Double Rainbow" video, viewed nearly 40 million times, to the quirky advertisement that launched the Dollar Shave Club business.[10] Videos are not the only media that go viral online. Photographs, GIFs, image memes, and even political or social causes—often anchored, certainly, by a particularly resonant image or video, as in the recent "Ice Bucket Challenge" videos in support of ALS research—can gain sudden attention, largely through the agency of social networks, and be widely and rapidly shared.

To cite a personal example, in January of 2013 my daughters Cadence (then twelve) and Emerson (then 9) posted a picture of themselves with their siblings Jonas and Rory (then 4) and Jude (then 1) on a Facebook page they titled "Twogirlsandapuppy." The children were holding a sign that read, "Hi World / We want a puppy! Our dad said we could get one if we get 1 million likes! So LIKE this!" with a note to the side that read, "He doesn't think we can do it!"[11] This picture quickly circulated through social media, first among friends and friends of friends in my and my wife's social networks—but soon far beyond. Within eight hours, the picture had received one million likes; within twenty-four hours, it had received nearly four million. In addition to these likes, the campaign's Facebook page was flooded with comments and private messages from people around the world who responded to the girls' campaign in (mostly) positive ways. These commenters cited a wide range of reasons for supporting the girls' viral plea, from a love of dogs, to religious identification (despite the picture having no overtly religious content), to shared Star Wars fandom, to—if I might indulge—the fundamental cuteness of my children. And of course, a great number of them cited a desire to prove a recalcitrant father wrong. In other words, the picture spread so widely and rapidly because it evoked many different points of identification among its viewers.[12]

Viral media isn't always so light or flippant, however, and it often seeks to inspire more serious forms of identification from those who view or read it. Consider, for instance, the "It Gets Better" campaign, inspired by a video created by columnist Dan Savage and his partner, Terry Miller, which organizers touted as "provid[ing] hope for lesbian, gay, bisexual, transgender and other bullied teens" through video messages submitted by hundreds of people.[13] These ranged from group posts by the employees of businesses, such as Facebook and Google, to videos by celebrities, such as Sarah Silverman and Stephen Colbert, to submissions from political figures, such as Secretary of State Hillary Clinton and President Barack Obama. While the effectiveness of such campaigns is a subject of significant debate, it is clear that they exemplify another kind of online viral media—politically activist rather than entertaining—that generates energy by encouraging viewers to share with their own social networks. The most successful viral media pieces, whether serious or light, become cultural touchstones in their own right, inspiring new content in response.

In *Going Viral*, Nahon and Hemsley use the term "viral events" to describe the constellations of engagement that emerge around viral media. A viral event includes not only an originary piece that is widely shared, but also the rich ecology of media that emerges around it—responses, reviews, remixes, mash-ups, and so forth. Consider, for instance, the hundreds of response and parody videos that appeared in the wake of "Gangnam Style," in which navy midshipmen, prisoners, lifeguards, moms and babies, and a wide variety of college groups imitated or reinterpreted scenes from the original video, sometimes with the original song as their background and sometimes with new lyrics that rework the song for specific communities of viewers.[14] These responses both testified to and contributed to the song's pervasiveness and cultural impact. The photograph, video, or text that spawns a viral event can quickly become subsumed by waves of new, responsorial media, linked by a particular artistic, aesthetic, or thematic idea.

Like my daughters' puppy campaign, the most successful viral phenomena provoke a range of reactions from their audience, provoking viewers or readers to respond and share widely, but for different reasons and toward different ends. Indeed, the "Twogirlsandapuppy" campaign was inspired when my eldest daughter saw a Facebook post in which two children held up a sign claiming their dad would get them a cat if they received one thousand likes. When she asked me whether she could do something similar for a dog, I demurred, claiming one thousand likes would be too easy, thus inspiring her to up the ante to one million. Her adaptation of the likes meme itself spawned a host of imitators, until "one million likes pleas" became internet clichés.[15] The most viral media are interpretively flexible and highly adaptable rhetorically, aesthetically, politically, or otherwise.

This notion of virality recalls the meme, first proposed by Richard Dawkins in *The Selfish Gene*, as a "replicator" for ideas, "a unit of cultural transmission . . . drifting clumsily about in its primeval soup" of "human culture." In Dawkins's original conception, memes describe far more than shared Internet content. Memes can be "tunes, ideas, catch-phrases, clothes fashions, ways of making pots or of building arches." For Dawkins the meme is not a metaphor, but a physical reality: "Just as genes propagate themselves in the gene pool by leaping from body to body via sperms or eggs, so memes propagate themselves in the meme pool by leaping from brain to brain via a process which, in the broad sense, can be called imitation."[16] While memetics has been increasingly marginalized as a scientific discipline, the idea of memes—the meme meme—continues to propagate and mutate with our culture. In particular, the word *meme* has shifted from referring to a generalized idea replicator and has instead become a shorthand term for viral content online, particularly for works that have become so widely known that an ecology of variations, remixes, and responses has emerged around the original pieces.[17] When discussing online content, then, the meme and virality are now entangled concepts—indeed, Dawkins's broader theorization of the meme has been largely replaced by discussions of content sharing and virality. As a way to consider the transmission of ideas, the meme proves useful, if only because it asks us to attend not to distinct cultural artifacts—this entire poem, that entire story—but to the more amorphous concepts and rhetorical figures that underlie specific artifacts and often propagate beyond them.

For critics, both the notion of memes and the viral metaphor are flawed because they occlude human agency, portraying the transmission of ideas as a deterministic, quasibiological process rather than as the result of individuals' tastes, choices, and social interactions. Henry Jenkins calls "viral media" a term "at once too encompassing and too limiting," conveying an idea of "circulation as the empty exchange of information stripped of context and meaning." He prefers the term "spreadable media," which in his view recognizes "acts of circulation as constituting bids for meaning and value" that "shape the cultural and political landscape in significant ways."[18] At present, however, "viral media" has resonated in cultural discourse in ways alternative terms have not. For practical reasons, we chose the name of the Viral Texts project in order to suggest analogies between the reprinted nineteenth-century newspaper and magazine snippets we are studying and more recent forms of media sharing online.[19] Not coincidentally, our use of this analogy has led to significant public interest in the project, which offers historical context for media outlets and others working to understand what had seemed to be a sudden and recent phenomenon. I would argue that such public engagement is in itself a significant good, particularly for literary-historical work in this current moment of "crisis"—whether real or imagined—in the humanities. In addition to these pragmatic motives, however, many of the models developed to describe online virality can illuminate rhetorical and systemic features of widely circulated nineteenth-century texts that are not readily apparent when described using established bibliographic or literary-historical models.

Going Viral in Nineteenth-Century Newspapers

Virality extends the notion of "the social text," first proposed by D. F. McKenzie in *Bibliography and the Sociology of Texts* and further developed by Jerome McGann through projects such as the Rossetti Archive. As McGann notes, when McKenzie first proposed the "social text" as an editorial procedure, prominent bibliographers "remarked that while McKenzie's ideas had a certain theoretical appeal, they could not be practically implemented" because "critical editing—as opposed to facsimile and diplomatic editing—was designed to investigate texts, which are linguistic forms, not books, which are social events."[20] In the Rossetti Archive, McGann seeks to represent the "social text" by including all editions of a given work in an online archive, rather than simply the "Reading Text" and "Variorum Text" of the standard critical edition. However, even the social text model remains focused on discrete works—books, most often, though also stories or poems—that can be collated and compared as distinct entities. Virality is messier, aligning fragmentary texts and textual echoes not only through books but also through ephemeral and hybrid media; the latter of these is exemplified by the nineteenth-century newspaper. The "viral text" of a particular poem would include official and unofficial reprintings, but also parodies, quotations, reviews, paraphrases, allusions, and more—what Julia Flanders has named "reception items." A theory of viral textuality must wrestle with unusually capacious ideas of "the text," including in its purview the continually shifting penumbrae of readers' responses that testify to that text's life within culture(s).

For this reason, virality proves especially useful for thinking about how texts circulated in the increasingly complex mass media ecology of the United States during the nineteenth century. During this time, newspapers and magazines proliferated, and this rapid expansion of the print sphere was accelerated by a system of content sharing among publications. The periodical press in the United States depended on "exchanges," through which editors subscribed to each other's publications (paying little to no postage for the privilege), and borrowed content promiscuously from each other's subscriptions.[21] Texts of all kinds were reprinted—typically without authors' or publishers' permission—across books, newspapers, and magazines. Content shared through the exchange system was not protected under intellectual property law. Instead, periodical texts were considered common property for reprinting, with or without modification—much as articles, music videos, and other content are shared online today among blogs and social media sites. And as is the case today, antebellum content creators reacted in disparate ways to these sharing practices. Some writers and editors compared reprinting to theft, decrying a system that popularized writers' work without supporting them financially. Others exploited the reprinting system in order to build a reputation that could be leveraged toward paid literary employment.[22]

The spread of "viral" content in nineteenth-century newspapers depended on a range of factors, from the choices of editors to the preferences of readers to the material requirements of composing a given day's issue. The frequently reprinted listicle "Editing a Paper," for instance, lays out the dilemma that faced nineteenth-century editors considering whether and how much to reprint:

> If we publish telegraph reports, people will say they are nothing but lies.
>
> If we omit them, they will say we have no enterprise, or suppress them for political effect . . .
>
> If we publish original matter, they find fault with us for not giving selections.
>
> If we publish selections, folks say we are lazy for not writing more and giving them what they have not read before in some other paper. (11 July 1863)[23]

The first reprinting of "Editing a Paper" identified by the Viral Texts project appears in the *Big Blue Union* of Marysville, Kansas, but even here an editorial preface claims that the list has been "going the rounds of the papers. If we knew in what paper it first appeared," the editor continues, "it would afford us pleasure to give the writer due credit." This piece and its preface illustrate much about editors' and, presumably, readers' attitudes toward reprinting, and how those attitudes might line up with modern ideas of viral media.

Considering nineteenth-century newspaper snippets as "viral media" allows us to frame their spread in terms of "rhetorical velocity," a term first developed by Jim Ridolfo and Dànielle Nicole DeVoss to describe online composition practices in which writers take reuse and remixing as a given and compose with an eye toward facilitating such reinterpretive acts. Such writers take as their primary assumption that a piece will be recomposed by others—reprinted or otherwise remediated. Ridolfo and DeVoss propose that "when academics uphold distinctions between author and producer, we are left in an uncomplicated, often acontextual space that does not provide the tools we need to best negotiate the ways in which production and authorship become more slippery in digital spaces and within remix projects." They argue, "The term *rhetorical velocity* means a conscious rhetorical concern for distance, travel, speed, and time, pertaining specifically to theorizing instances of strategic appropriation by a third party."[24] In other words, "rhetorical velocity" posits "the text" through multiple dimensions, charting its uses and movements—both social and geographic—alongside its evolving content. What's more, a piece need not be *consciously* crafted for a wide audience to have rhetorical velocity; if it is compelling, concise, and easily modified, then it can go viral with or without its creator's knowledge.

While Ridolfo and DeVoss refer specifically to composing practices online, the frame of rhetorical velocity offers insight into widely reprinted newspaper content during the nineteenth century. Nineteenth-century editors relied on the exchange system to provide engaging content, and they in turn composed (or solicited) original pieces with an eye toward their readers *and* those of the papers with which they exchanged. In the first post–Civil War issue of the *Pulaski Citizen*, for instance, editor Luther W. McCord apologizes for the sorry state of "The News" in the paper because "we have no exchanges yet, from which to make up our news items. Our readers can readily appreciate," the squib continues, "the impossibility of making an interesting paper without something to make it of." McCord then assures readers that they "hope to have a full list of exchanges by next week and, per consequence, a more readable number of the *Citizen*" (January 5, 1866). This apology echoes a common notion among editors in the period: newspapers that aggregated content from exchanges were of higher and more consistent quality than newspapers written entirely by locals. In other words, McCord assumes that his primary job will be selecting and propagating writing from elsewhere—contributing to the rhetorical velocity of content written for a distributed network, not for individual newspapers.

We must therefore assume that newspaper editors and writers were concerned with the rhetorical velocity of what they published; a newspaper whose content was regularly reprinted in other newspapers would, presumably, soon be added to more exchanges, as editors further down the line sought the source of the pieces they encountered in intermediary papers. This would, in turn, increase the popular newspaper's circulation and subscription fees. Indeed, when considering nineteenth-century newspaper snippets, we might speak of "composing for recomposition" in a more technical way, using "composition" not only in its modern sense, as a near synonym for "writing," but also as a printers' term of art. As scholars such as Ellen Gruber Garvey have shown, texts were reprinted in newspapers to help editors compose entire daily or weekly newspapers with small staffs. "By yoking together scattered producers who shared labor and resources by sending their products to one another for free use," the network of newspapers sustained the proliferation of its medium.[25] In other words, reprinting existed in large part to meet the material needs of publication. Many of the changes introduced into texts as they circulated through the newspaper network—a line removed here, two lines added there—were motivated by these practical considerations, as a given newspaper's compositors shaped exchange content to fill empty spaces on a nearly composed page. It seems reasonable to presume that as a newspaper's compositors prepared their pages each day or week, they expected—perhaps even hoped—that other compositors in their exchange networks would later recompose their texts, extending the texts' rhetorical velocity to reach distant audiences.

As with modern viral media online, one of the clearest testaments to the rhetorical velocity of a text in the nineteenth century was the emergence of parodies. A parody assumes widespread audience familiarity with the original piece it mocks (or enlists in the service of mockery)—otherwise much of its humor would be lost on readers. In rewriting the piece for comedic or satirical effect, the parody is both a distinct bibliographic artifact and part of the viral event surrounding the text it parodies. Consider, for instance, the poem "The Inquiry," by Scottish poet Charles MacKay.[26] The original text meditates,

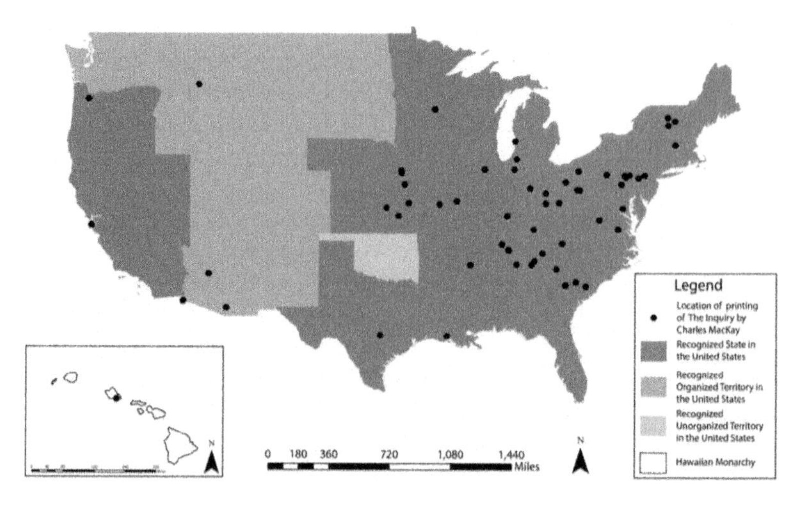

Figure 2.2 Reprintings of Charles MacKay's "The Inquiry" in nineteenth-century American newspapers. The state boundaries here are from 1870, about midway through the text's life in the press, though these boundaries shifted considerably during the decades the text circulated. This map was prepared by Viral Texts project intern Laura Eckstein, an undergraduate at Haverford College.

through four stanzas, on the difficulties of earthly life. In each of the first three stanzas, the speaker questions the "winged winds," "mighty deep," and "serenest moon," asking whether there is on earth "some spot / Where mortals weep no more" or "Some valley in the west . . . free from toil and pain" where the "weary soul may rest." To these and similar questions, nature answers "No!" in each stanza's closing couplet. Finally, the speaker asks "my secret soul," "Is there no happy spot / Where mortals may be bless'd, / Where grief may find a balm / And weariness a rest?" To this final inquiry, the speaker receives a new response: "Faith, Hope, and Love, best boons to mortals given / Wav'd their bright wings, and whisper'd—'Yes, in Heaven.'"

MacKay's sincere affirmation of mainstream Christian ideas of the afterlife was widely loved and shared, appearing not only in many newspapers, but also in many nineteenth- and early-twentieth-century poetry anthologies. The poem also sold well set to sheet music.[27] According to many of his biographers, "The Inquiry" was a particular favorite of Abraham Lincoln, who may have purchased the sheet music for the piece during his time in Springfield.[28] But the poem also provided rich fodder for satirists, who copied its basic structure and many of its central images but recomposed its details for comedic effect. On March 4, 1857, the *Grand River Times* (Grand

Haven, Michigan) printed "A Parody. By an Old Bachelor," which reworks MacKay's devotional lines into a bachelor's misogynistic lament. This speaker inquires also of the wind, the waves, and the moon in turn, but he asks if there is "some spot" where, for instance, "women fret no more," "babies never yell," "hoops are out of place," "muslin is not known," or "weary men may find / A place to smoke in peace." As in MacKay's original, all three forces rebuff the speaker, though this time with colloquialisms: "Nary place," "Yeou git eout," and "Pooh!" And the parody borrows the final couplet of MacKay's poem almost exactly, as the speaker learns from Faith, Hope, and Truth—not Love, perhaps appropriately—that "females never go" to one place that "bachelors are blessed" and "may dwell in peace": "*in Heaven!*" Here the conventional piety of MacKay's poem becomes a punchline about domestic tranquility, or the lack thereof; this parody too "went the rounds" of the papers.

This first parody was not the last recomposition of "The Inquiry." On June 6, 1857, the *Keowee Courier* (Pickens Court House, South Carolina) printed "Parody Parodied," which the editors introduce as written by "some fair writer" who "thus retaliates on the parodist whose production we published some days since." The editors continue, insisting that "both sides must be heard, and we give the lady a chance." In this poem, the speaker again queries the elements, wondering if a place exists where "bachelors come no more," "no moustache is seen," and "cigars are not," a place where "weary girls may find / A rest from soft dough faces" continually wooing them. Again after three denials—one tinged with temperance rhetoric, as the sea murmurs, "Not while brandy smashes live"—the three angels of the final couplet assure the speaker that she will find rest from men in heaven. This parody is twice removed from MacKay's original poem; it is a recomposition of a recomposition. Nevertheless, it clearly belongs to the same viral event as "The Inquiry," both exploiting and feeding back into the rhetorical velocity of MacKay's piece.

This viral event was in fact so widespread that the central images and structure of "The Inquiry" would eventually be invoked without any direct reference to MacKay's original, and even without the visual markers of poetry. In November 1857, for instance, the *Raftsman's Journal* (Clearfield, Pennsylvania) first printed this satirical squib, as a prose block rather than in poetic lines:

> Query.—Tell me ye winged winds that round my pathway roar, do ye not no [sic] some quiet spot where hoops are worn no more? Some lone and silent dell, some Island or some cave, where women can walk

> three abreast along the village pave? The loud winds hissed around my face, and snickering answered, "nary place."[29]

This is the entire piece, and we might imagine it as written and then reprinted primarily to help compose the page—to fill a small gap. That MacKay's poem was chosen for this short joke, however, tells us much about the extent of the text's virality, the practical force of its rhetorical velocity. Editors who printed this squib in their newspapers expected readers to get the joke, quickly and without additional context. Moreover, they could make the joke in shorthand, offering what is in effect a one-paragraph prose poem in place of the four stanzas of the original text or its first parodies. Clearly "The Inquiry" had gone viral to such an extent that an offhand, even oblique, reference could be expected to resonate with readers. Like a ubiquitous online meme today, "The Inquiry" was so familiar that recomposing it became a performance of simultaneous affection for and ironic distance from the original work.

Despite these transhistorical resemblances, rhetorical velocity remains a difficult measure of comparison between the early nineteenth century and the early twenty-first. The particular metaphor of virality emerged in large part because digital platforms enable observers to track the spread of content on an individual level (whether through views, likes, or retweets)—a granular view of information flow that seemed comparable to the spread of disease from person to person. Modern content-tracking systems allow us to see that a given YouTube video was viewed two million times, liked on Facebook one million times, tweeted another five hundred thousand times, and so forth. The idea of virality inheres less in the kind of content that is shared online, then, than in the way the online medium lets us observe and track the social processes of sharing. By contrast, while we can know that "Editing a Paper" was printed in at least 134 newspapers, we cannot know how many readers actually read it, much less how many valued it.

Of course, even modern, granular measures of views and likes do not tell us that much about how viral media content is received and used. Writing about likes on Facebook, for instance, Robert Payne notes:

> Clicking "Like" may not only mean "I like this" or "I enjoyed this item" but a whole host of possible responses, including "I think this is amazing," "I think this is reasonably interesting," "I think this is stupid," "I want to increase traffic to this page" and "my friend is

> pressuring me to Like her page and I don't really like it but I like her so what else can I do?". In some circumstances, given the lack of alternatives, "Like" may even mean "Dislike" . . . all of these motivations for clicking "Like"—and all of them are valuable—lead to the same result of the item increasing in circulation and possibly going viral, but nothing of these outcomes registers the multiplicity of user engagements. "Like" flattens out this multiplicity and becomes more about technical functionality, an unreflexive bodily response, a way of interacting with the media object without having to express individual motivation.[30]

Payne describes the like button as a flat signal, attesting to nothing more than the bare fact that a picture, text, video, or other media was clicked after a binary—and perhaps arbitrary—decision. "Like" can even mean its opposite in many cases, given users' desire to signal a response to content when only one signaling mechanism is provided. In other words, the mass of potential reasons motivating someone to click "like" is buried by the functionality of the platform.

Like a modern like or retweet, the reprinting of texts in nineteenth-century American newspapers is a flat signifier. Editors selected texts from other publications for a wide range of reasons, from admiration to repudiation to capitulation. Editorial paratexts explaining why a given text was reprinted are relatively rare; the occasional line of explanation proves immensely valuable.[31] Such commentaries on reprinted content sometimes praise it, as when the *Grand River Times* (Grand Haven, Michigan) "commend[s] to all our patrons and friends" a list of rules for home education "for their excellence, brevity, and practical utility."[32] The *Grand River Times*'s editorial endorsement of this listicle builds on—and includes as part of its reprinting—the *Plymouth Pilot*'s (Plymouth, Indiana) declaration that the list was "worthy of being printed in letters of gold, and placed in a conspicuous position in every man's household."[33] These and similarly laudatory sentiments are the most common editorial comments on viral texts in the period. As we have seen, editors touted reprinting as a mechanism for improving their newspapers and so were very likely to praise reprinted content as entertaining, enlightening, or useful for their readers. But editors sometimes inserted a line indicating their skepticism or hostility toward republished pieces. In these cases, the text is reprinted so that its errors can be more widely known and condemned.

Whether reprinting to praise or condemn, editors consistently invoked their readership as the inspiration for their selections, claiming to reprint content needed, solicited, or even created by readers. In an article often titled "How to Do Up Shirt Bosoms," for instance, the *Democrat and Sentinel* (Ebensburg, Pennsylvania) noted that "we often hear ladies expressing a desire to know by what process the gloss on new lines, shirt bosoms, &c., is produced, and in order to gratify them we subjoin the following recipe."[34] Here the editors claim to be printing in direct response to their readers' expressed needs. This language of needs met and interest sated—of *usefulness*—pervades reprint culture, aligning viral newspaper literature with what Franco Moretti calls "the style of the useful" in nineteenth-century bourgeois writing.[35]

Many of these useful texts were even penned, at least in part, by readers rather than editors. A widely reprinted piece on the health benefits of tomatoes began as an editorial response to an inquiry from one "P. B. T."; this individual's letter, appearing in the *Edgefield Advertiser* (Edgefield, South Carolina), notes that "so little attention has been given to [tomatoes'] preservation" that many Americans are unaware they can be eaten year-round and report finding their taste unpleasant. P. B. T. asks the editors to correct these myths, as tomatoes' "presence on the table at any, or even with all meals of the day, is quite acceptable. A notice from you at this time," this correspondent continues, "would be of service to at least one of your readers." The *Advertiser* obliges, following P. B. T.'s letter with a long "Answer by the Editor" concurring that "the Tomato has long been known and used for culinary purposes in many portions of Europe . . . and within a few years has become a general favourite in this country." The piece then cites the many health benefits of the fruit, each ascribed to a particular doctor or professor, before offering six recipes for cooking or preserving tomatoes. In later reprintings, P. B. T.'s letter largely disappears, summarized in a single introductory sentence—"To many persons there is something unpleasant, not to say disgusting, in the flavor of this excellent fruit"—which is then followed by some version of the original editor's reply, though sans the "Answer by the Editor" subheadline of the original article.[36]

Reprinted pieces in nineteenth-century newspapers were in fact often cocreated by correspondents and editors, as the newspapers' readers and subscribers wrote through the medium to the wider community it created and sustained. The correspondent who sparked "The Tomato" seeks to correct a perceived error in his or her community and sees the newspaper as a vehicle for a public address. In such pieces we might recognize something more akin to modern viral media, in which the readers drive the popularity of content from the bottom up, rather than editors determining what readers need from the top down. Indeed, we might even identify readers corresponding with their newspapers as a form of social media, an attempt to connect private practice with a broader community.

Nevertheless, in this system editors ultimately controlled what did and did not appear in the newspaper. This model, of editors swapping texts through semiformalized and formalized exchange networks, might seem too organized to fit our expectations for viral media. Such a distinction, however, stems from a common misconception about modern viral media online, one that imagines virality to be driven largely or entirely by the crowd. This is often framed in contrast to "traditional" media, in which gatekeepers—editors, journalists, writers, reporters—choose what information is worthy of distribution. This idea of virality sees gatekeepers as removed entirely from the information loop and casts online content not as a "one to many" medium, but instead purely as a "many to many" medium. Such a frame is difficult to reconcile with nineteenth-century exchange reprinting.

However, viral media researchers are increasingly finding both top-down and bottom-up models insufficient for describing the way content actually goes viral online. Instead of discovering sudden and unprompted swells of attention from individuals, they note the importance of gatekeepers—though, to be fair, not necessarily the *same* gatekeepers as in twentieth-century print economies—in helping a given piece gain an initial audience who will spark its viral spread. Nahon and Hemsley describe viral events as driven by a complex interplay of "organic" sharing among social media users, coverage by mainstream media sources, and attention from "opinion leaders"—such as the "about 20,000 elite users" who "attract about 50 percent of the attention (measured by the spread of URLs) on Twitter," which currently claims about 270 million active monthly users.[37] Consider the "Double Rainbow" video, which has garnered nearly 40 million views on YouTube. In many ways it epitomizes our ideas about Internet virality—it's an amateur video, shot without any ambitions regarding widespread use or popularity, that people found remarkable for the filmmaker's rapturous narration. However, the video was originally posted on January 8, 2010, and it sat largely *un*remarked on YouTube until late-night comedian Jimmy Kimmel tweeted about it nearly seven months later, on July 3, 2010. The widely connected Kimmel—a gatekeeper, in other words—seeded the video's viral success, bringing the piece to the attention of a large number of people who would otherwise have been unaware of its existence and who then shared it with their smaller but numerous networks.

Like these modern gatekeepers, nineteenth-century newspaper editors served as catalysts for virality. When one newspaper reprinted a piece, it went out not only to their readers but also to all the editors of other newspapers with whom they exchanged. The fact that one newspaper chose to reprint a text boosted its signal for editors down the line; when many newspapers did so, subsequent editors would often note that the selection was "going the rounds" of the papers. A piece "going the rounds" had, we might say, achieved substantial rhetorical velocity. Added to its intrinsically interesting features was the fact of its reprinting, of its virality. Like modern viral media online, nineteenth-century viral texts propagated, at least in part, due to network effects.

Nineteenth-Century Newspaper Networks

The practice of reprinting illuminates the social, technical, and business networks that underlay nineteenth-century newspaper production. As Payne argues about advertising that goes viral online, there is a "paradox at the heart of the relationship between content, structure and participation that vexes an easy understanding of popularity" on social media platforms. "The specific content of a given item," he notes, "may account for its viral success *at the same time* as having little or nothing to do with the successful functioning of social networks as distributors of content."[38] While we might read widely reprinted newspaper snippets as signals of cultural priorities and diversions, then—as I have done and will do in future—we can also use them to trace the broader contexts that enabled and even required their circulation. As with viral content online today, the rhetorical velocity of nineteenth-century newspaper selections depended on a mixture of formal and informal networks of influence, which we can begin to recover through the texts themselves. The bibliographies of viral pieces, gathered en masse, offer new purchase on the sociology of texts during the nineteenth century.

There are few clearer signals of influence among newspapers than shared reprinting, particularly when considered at scale. That two newspapers printed a few texts in common tells us little about their relationship; that two newspapers printed hundreds or thousands of texts in common, however, suggests a connection worth closer attention. At the other end of the scale, the degree to which individual texts saturate a given network can help us understand those texts' circulation and their significance. Considering reprints as signals of social interconnection among publications, weak in the singular but powerful in the aggregate, we can use shared texts to graph conversation and influence across nineteenth-century periodical culture—much as we can map conversation and influence in a Twitter network based on replies and retweets. Unlike Twitter, of course, our archive of nineteenth-century periodicals remains far from comprehensive, which means that counting reprints can only take us so far toward understanding the influence of a given text.

Focusing on raw numbers of reprints can in fact elide the circulation of texts through smaller geographic, ideological, religious, or political networks. Virality is a model that can apply (or not) at different scales, so we might talk of one poem "going viral" across the entire system of nineteenth-century print culture, while the transcription of a political speech "went viral" among the newspapers of one political party only. In this way, virality becomes not simply a measure of raw popularity, but instead a frame for comparison among texts and publications. Much like social interactions online, historical "viral texts" can reveal relationships or influence among publications. At scale, such texts can be used to reconstruct the networks of nineteenth-century print. They can show how information circulated among publications, which publications most influenced (or were most influenced by) which others, and how various communities were constructed through textual exchange.

These are some of the major goals of the Viral Texts project. Though this work is nascent, it has already suggested new ideas about American print, particularly in the period before the Civil War.[39] To cite one small but telling example of how scale might shift scholarly attention, our network analyses of reprinting have pointed toward newspapers in understudied cities in the South and Midwest as more central to textual exchange during the period than previous criticism has recognized. Scholars interested in nineteenth-century periodicals have often focused on large-circulation newspapers from publishing centers, such as New York, Boston, and Philadelphia—by necessity, as these papers are the most easily accessible in frequently consulted archives. However, when our data is modeled as a network (as you can see at http://networks.viraltexts.org/1836to1860/index.html), it is striking how papers such as the *Nashville Union*, *Sunbury American* (Sunbury, Pennsylvania), *Daily Dispatch* (Richmond, Virginia), and *Fremont Journal* (Fremont, Ohio) emerge as shaping influences on the wider network of newspapers outside the urban Northeast.[40] These early findings suggest that influence across the nineteenth-century print network was far more distributed than scholars have typically assumed, a theory we will continue to test as we incorporate more data into our study.

For example, while the 1836–60 graph shows the *New York Tribune* and *Evening Star* (Washington, DC) as influential nodes in the network of nineteenth-century reprinting, their degrees are roughly comparable to those of the *Ottawa Free Trader* (Ottawa, Illinois), *Daily Dispatch* (Richmond, Virginia), and *Nashville Union*, followed closely by a range of regional papers, such as the *Glasgow Weekly Times* (Glasgow, Missouri), *Edgefield Advertiser* (Edgefield, South Carolina), *Raftsman's Journal* (Clearfield, Pennsylvania), and *Burlington Free Press* (Burlington, Vermont). This effect certainly stems in part from the data we are analyzing—Chronicling America's holdings skew disproportionately toward smaller and more rural papers—but it nonetheless provides a valuable window into textual exchange outside the urban centers that often monopolize historical and literary criticism. For the study of virality, the dispersed regionalism of this archive might be as much a boon as a hindrance; for a single text to saturate the newspapers that comprise Chronicling America, it had to resonate in large cities and small villages, in the North, South, East, and West.

Moreover, this network view of textual sharing opens up new possibilities for comparison between and among individual viral texts. Consider, for instance, Charles MacKay's religious poem "The Inquiry" and "Interesting Statistics," a column of factoids supposedly compiled by "A gentleman claiming to be a 'friend of the human race,'" including statistics about the "whole number of languages spoken in the world," human life expectancy, religious profession, and marriage. "The Inquiry" (graph available at http://networks.viraltexts.org/1836to1860-Inquiry/index.html) and "Interesting Statistics" (graph available at http://networks.viraltexts.org/1836to1860-InterestingStats/index.html) were printed a comparable number of times prior to 1861: we have identified 33 reprints of the former and 32 of the latter. To put it another way, the pieces were both printed in about one quarter of the newspapers in our study for the antebellum period.[41] Though they were both reprinted in certain newspapers, however, such as the *Evening Star*, *Nashville Union*, and *Ottawa Free Trader*, their networks were also in many ways distinctive. "Interesting Statistics," for instance, was printed in several highly influential newspapers that missed "The Inquiry," including the *New York Daily Tribune*, the *Rutland Herald*, and the *Fremont Journal*. While "The Inquiry" was also printed in influential papers, those that printed only "The Inquiry" are largely concentrated in one state: the *Vermont Watchman*, the *Burlington Free Press*, or the *Middlebury People's Press*.

Overall, "Interesting Statistics" was reprinted in more of the newspapers identified by our study as central to textual exchange during the period. Thus we might expect to see this piece printed more often than its counterpart, as more newspapers drew from their network's influencers. And looking at a larger Viral Texts data set that includes newspapers after 1861, the rhetorical velocity of "Interesting Statistics" does seem to have accelerated at a higher rate—though admittedly only slightly higher—than that of "The Inquiry." The former was ultimately reprinted at least 120 times through the nineteenth century, while we have identified only 93 reprintings of the latter during that same period. Of course, as a poem, "The Inquiry" was also featured in other media not included in this study, including poetry anthologies and sheet music. It seems likely that the overall cultural impact of "The Inquiry" was greater than a more ephemeral trivia column. Nevertheless, a network effect is discernible for the two pieces within the system of nineteenth-century newspapers. The piece that found purchase in the most influential newspapers was ultimately reprinted in more newspapers—propagated, as Payne argues about Facebook likes, less by the inherent value of the content than by the features of the platform itself, by the medium.

These analyses are only a starting point for considering the system of nineteenth-century print as a network. Significant work will be necessary both to create more sophisticated models of publications' connections and to analyze texts not only across the total system, but also within smaller contexts. In the Viral Texts project, we are working toward more nuanced investigations that take into account both genre and topic (for the texts) and affiliation (for the newspapers). We are currently annotating our initial results to this end, marking the topics and genres of the most widely reprinted texts we have thus far identified (e.g., temperance or abolition, poetry or political speech), as well as the shifting editorial, political, social, and religious affiliations of the newspapers in our study. These annotations will allow us to better compare the rhetorical velocities of viral texts, discovering whether poetry, for instance, was more likely to be reprinted by religious papers than by secular ones, or revealing the degree of cross printing between ideologically opposed newspapers. We are also conscious that the national limitations of our current study do not adequately account for the transatlantic print networks that brought texts like "The Inquiry" to the United States and exported American texts, as Bob Nicholson demonstrates in his study of American jokes in the British press.[42] We are currently exploring partnerships that will allow us to include British and other newspapers in our study to further explore transnational reprinting. Even as we pursue these larger questions, however, our early experiments demonstrate the potential for modern social-network analysis to illuminate aspects of nineteenth-century print culture that have been largely obscured by the scale of the newspaper archive. If virality both creates and is created by its platform and by that platform's gatekeepers, then a granular study of viral texts in the nineteenth century will bring both the newspaper medium and its messages into sharper focus.

CONCLUSION

The network graph allows us to map not geographic space, but social space: to visualize the social lives of texts, their viral spread through the interrelated operations of people and platforms. As models, they are partial, privileging certain connections—in this case, shared reprints between publications—and obscuring others. As Franco Moretti argues of geographic maps, the network graph is "a new, *artificial* object" that "possess[es] 'emerging' qualities which were not visible at the lower level."[43] In this case, such models can help us grapple with the messy, energetic, complex system of newspaper exchanges in the nineteenth-century United States, both tracing the rhetorical velocity of individual snippets across the system and using snippets en masse to envision a larger network structure. What emerges from such modeling is a more capacious understanding of textual influence, including in its purview not only the major print centers of the period but also the regional publications that comprised the bulk of nineteenth-century newspaper culture.

As texts moved through newspaper networks, they assumed new forms and sometimes sparked wider viral events—flurries of responsorial textual activity. The metaphor of virality, while imperfect for describing the circulation of cultural ideas and artifacts, nonetheless proves useful for framing the social text: its vectors of transmission, its diverse audiences, its many and varied modes of expression. Perhaps most importantly, virality focuses exclusively on cultural repetition and remediation, on modeling the social diffusion of content. As I wrote in the introduction to this essay, these borrowed terms—*virality*, *memes*, *viral events*, *rhetorical velocity*—are in essence new ways of framing bibliographic description (what was printed, where, and when?). But virality's overarching theme remains contagion; the metaphor is somewhat unsatisfying because it implies a paucity of human agency, the unintended spread of disease rather than the passing on of useful, beautiful, or entertaining content. This fact, too, perhaps explains the pervasiveness of virality for describing online content, which is often dismissed as trite and ephemeral. Better, then, that it should spread by contact than by choice.

But this caveat also points to the central reason the virality metaphor is ultimately useful for thinking in new ways, bibliographical and literary-historical, about nineteenth-century print culture. Considering virality forces us to subject textual transmission, repetition, and circulation to systematic scrutiny, to look beyond *only* human choice and examine the technological, economic, and network effects that shaped what texts nineteenth-century readers encountered in their newspapers. The newspaper was a truly mass medium, both in its reception and in its creation. Editors sifted through a mass of content to choose the selections they would distribute to their readers, who included other editors, who would in turn sift and select. To achieve the rhetorical velocity required to "go viral" within this system, a given piece had to both resonate with readers and fit the material requirements of composition and recomposition. The model of the public sphere has been used to describe the workings of print during the nineteenth century; virality offers a complementary model that requires consideration of people and platform, message and medium, for understanding how ideas circulated during the period.

NOTES

1. This is the earliest printing of this piece that I have discovered; there is nothing in the *Ashtabula Weekly Telegraph* issue pointing to the provenance of the piece.
2. I write more about the listicle as a genre in another forthcoming article. As I say in that piece, it is difficult to find academic citations for the listicle, but Arika Okrent defines the genre succinctly for *The University of Chicago Magazine* (January/February 2014) in "The Listicle as Literary Form": "A listicle is an article in the form of a list." Okrent then attempts to account for the dominance of the listicle online and concludes by defining a listicle in the listicle form:

   Eight fun facts about the listicle

   1. A listicle is an article in the form of a list.
   2. It is kind of like a haiku or a limerick.
   3. It has comforting structure.
   4. It makes pieces.
   5. It puts them in an order.
   6. Language does that too.
   7. Sometimes with great difficulty.
   8. Lists make it look easier.
3. This figure, as will be described below, is based on computational analysis of newspapers in the Library of Congress's open-access Chronicling America historical newspaper archive and Making of America historical magazine and journal archives. Based on our tests across commercial

archives of nineteenth-century periodicals, we would expect to find many more reprints of this piece in collections managed by Readex, ProQuest, or EBSCO.

4. The reprinting immediately preceding *The Appeal*'s, in the March 1, 1899, *Somerset Herald*, included the even more blunt title "You Are Wrong."
5. For more on the aims and progress of the Viral Texts project, see http://viraltexts.org.
6. For a detailed description of how the project algorithm works, see David A. Smith, Ryan Cordell, and Elizabeth Maddock Dillon, "Infectious Texts: Modeling Text Reuse in Nineteenth-Century Newspapers," in *Proceedings of the Workshop on Big Humanities* (IEEE Computer Society Press 2013). This version of the algorithm looked for all reprinted texts in the corpus from its earliest issues (around 1835) to 1870, and then looked for additional reprintings of identified texts from 1870 to the end of the nineteenth century. In other words, the only texts from the period 1870–99 included here are those with a first reprinting before 1870. We chose this cutoff not as a strict periodizing marker, but instead as a way to keep computational needs manageable as we scale up the project.

   It is difficult to say with absolute certainty what percent of the newspapers in our corpus reprinted a given text, as the raw number of newspaper titles in Chronicling America does not accurately reflect the actual number of distinct newspapers in the corpus—often a single newspaper changed titles several times during its run, but each title is listed as a separate newspaper in the CA metadata. One of our tasks in the Viral Texts project has been to merge instances of "the same" newspaper in our database so that we can more accurately assess how thoroughly a given text saturated the network of newspapers. We have mostly accomplished this goal for pre-1861 newspapers but are still working through the task for post-1861 publications. In any case, "Follies" was reprinted in at least 136 of 712 (unmerged) newspapers. I use the formulation "at least XX newspapers" throughout this essay because we can be confident about those reprints we have identified, as I have checked each of the clusters cited here thoroughly. We cannot know, however, how many reprints our algorithm missed in our current corpus, due to very poor OCR or an odd selection process by a given editor, and we cannot know how many reprints exist in undigitized newspapers or in other, closed online corpora. As we have discovered through targeted searches in commercial archives (to which we do not have data-level access for text mining), our findings seem strongly indicative of larger reprinting trends.
7. J. Gerald Kennedy, "Introduction," *Poe and the Remapping of Antebellum Print Culture*, eds. Kennedy and Jerome McGann (Baton Rouge: Louisiana State University Press, 2012), 4.
8. Karine Nahon and Jeff Hemsley, *Going Viral* (Malden, MA: Polity Press, 2013), 16.
9. The video remains the most viewed item on YouTube and is fast approaching two billion views as of October 2013. Psy, "Gangnam Style," http://www.youtube.com/watch?v=9bZkp7q19f0.

10. The "Double Rainbow" video was uploaded to http://www.youtube.com/watch?v=OQSNhk5ICTI on January 8, 2010. Dollar Shave Club's "Our Blades Are F***ing Great" was posted to http://www.youtube.com/watch?v=ZUG9qYTJMsI on March 6, 2012, and has been viewed nearly 15 million times.
11. This page, including its original picture and many following, can be found at https://www.facebook.com/Twogirlsandapuppy. The campaign also received wide media coverage, including several outlets interested in how we might contextualize this event within a wider understanding of viral media. See Rebecca J. Rosen's *Atlantic* piece, "The Viral-Media Prof Whose Kids Got 1 Million Facebook Likes (and a Puppy)" (http://www.theatlantic.com/technology/archive/2013/01/the-viral-media-prof-whose-kids-got-1-million-facebook-likes-and-a-puppy/267338/), Radio Boston's "From Hawthorne to Facebook: How One Social Media Scholar Got Schooled" (http://radioboston.wbur.org/2013/01/25/social), or this interview on CBC's *Q* (http://www.cbc.ca/player/Radio/ID/2330327663/).
12. Readers, we got them a puppy: http://milliesmillions.ryancordell.org/.
13. See http://www.itgetsbetter.org/ for more about the "It Gets Better" project.
14. That these responses often parody a video that is itself already parodic gestures toward many potential conversations about cultural translation and virality for which I unfortunately have little space here.
15. One can read all about the origins and life of these "one million likes pleas," including a synopsis of my daughters' campaign, at the Know Your Meme website, http://knowyourmeme.com/memes/one-million-facebook-likes-pleas.
16. Richard Dawkins, *The Selfish Gene*, 30th Anniversary Edition (Oxford: Oxford Univ. Press, 2006), 192.
17. The Know Your Meme website (http://knowyourmeme.com/about) defines the "Internet meme" as "a piece of content or an idea that's passed from person to person, changing and evolving along the way. A piece of content that is passed from person to person, but does not evolve or change during the transmission process is considered viral content."
18. Henry Jenkins and Sam Ford, *Spreadable Media: Creating Value and Meaning in a Networked Culture*, Kindle Edition (New York: NYU Press, 2013), 20, 44, 194.
19. Media outlets, such as *Wired* magazine, *Lapham's Quarterly*, *Slate*'s The Vault blog, and NPR's *On the Media*, have covered the project, in large part because the media is hungry to understand modern virality online.
20. Jerome McGann, "From Text to Work: Digital Tools and the Emergence of the Social Text," *Romanticism on the Net* 41–42 (2006), http://www.erudit.org/revue/ron/2006/v/n41-42/013153ar.html.
21. For more about newspaper exchanges and nineteenth-century reprinting, see the introduction to Candy Gunther Brown's *The Word in the World* (University of North Carolina Press, 2004); Kyle Robert's "Locating Popular Religion in the Evangelical Tract: The Roots and Routes of

the Dairyman's Daughter," *Early American Studies* (2006); Meredith McGill's *American Literature and the Culture of Reprinting, 1834–1853* (University of Pennsylvania Press, 2007); David M. Henkin's *The Postal Age* (University of Chicago Press, 2007); Leslee Thorne-Murphy's "Re-Authorship: Authoring, Editing, and Coauthoring the Transatlantic Publications of Charlotte M. Yonge's *Aunt Charlotte's Stories of Bible History*," *Book History* (2010); Ellen Gruber Garvey's *Writing with Scissors: American Scrapbooks from the Civil War to the Harlem Renaissance* (Oxford University Press, 2012); Bob Nicholson's "'You Kick the Bucket; We Do the Rest!': Jokes and the Culture of Reprinting in the Transatlantic Press," *Journal of Victorian Culture* (2012); Ryan Cordell's "'Taken Possession of': The Reprinting and Reauthorship of Hawthorne's 'Celestial Railroad' in the Antebellum Religious Press," *Digital Humanities Quarterly* 7.1 (2013); and Will Slauter's "Understanding the Lack of Copyright for Journalism in Eighteenth-Century Britain," *Book History* (2013). The 2013 *Book History* also includes Meredith McGill's overview of scholarship on copyright: "Copyright and Intellectual Property: The State of the Discipline."

22. In "Reprinting, Circulation, and the Network Author in Antebellum Newspapers" (forthcoming in the August 2015 issue of *American Literary History*), I contend that authorship is at best an amorphous configuration in a print culture founded on sharing, exchange, and remediation of texts, the majority of which were anonymously authored. Often the accomplishment or creative act within the antebellum culture of reprinting hinged less on original composition than on savvy curation. Editors prided themselves on finding and reprinting pieces their readers would admire. For those readers, the newspaper system itself filled the authorial role. For widely reprinted snippets, textual authority and literary value were vested not in the genius of their author, but instead in their circulation and perceived usefulness.
23. First identified in our study in the *Big Blue Union* of Marysville, Kansas, and reprinted in at least 134 other newspapers between July 1863 and July 1897, with the bulk of reprintings in the 1860s and 1870s.
24. Jim Ridolfo and Dànielle Nicole DeVoss, "Composing for Recomposition: Rhetorical Velocity and Delivery," *Kairos* 13 (2, 2009) http://kairos.technorhetoric.net/13.2/topoi/ridolfo_devoss/index.html
25. Garvey, *Writing with Scissors* 31.
26. First uncovered in our study in the *Illinois Free Trader* of November 27, 1840, the "viral event" of this poem includes at least 93 texts in American newspapers between then and July 7, 1899. The poem was printed under many titles, the most common of which, besides "The Inquiry," is the first line, "Tell Me, Ye Winged Winds." It was also collected in many poetry anthologies throughout the nineteenth and early twentieth centuries.
27. An 1841 book of sheet music for the poem can be found in the Library of Congress's *Performing Arts Encyclopedia* at http://lcweb2.loc.gov/diglib/ihas/loc.music.sm1841.380450/default.html. Later editions of

the music can also be found in the Library of Congress's digital collections. The musical version of the poem testifies to its wide cultural impact—or, in the spirit of this essay, to its virality.

28. This claim is most recently made in Michael Burlingame's *Abraham Lincoln: A Life* (Baltimore: Johns Hopkins University Press, 2008). In Ellis Paxson Oberholtzer's 1904 biography, *Abraham Lincoln* (Philadelphia: George W. Jacobs and Company), Oberholtzer calls "The Inquiry" a "fugitive poem" before noting, "Ballads gleaned from newspapers and accidental sources lingered in [Lincoln's] mind if they touched some sad, responsive chord in his nature" (342). The claim that Lincoln purchased the sheet music of the poem can be found in Philip D. Jordan, "Some Lincoln and Civil War Songs," *The Abraham Lincoln Quarterly* (September 1942), 138.
29. First identified in our study in the November 5, 1857 issue of the *Raftsman's Journal*, this squib was itself reprinted in a number of other newspapers.
30. Robert Payne, "Virality 2.0," *Cultural Studies* 27 (4, 2013): 549.
31. This is comparable, perhaps, to the few words of commentary that sometimes accompany a retweet and that offer insight into a user's motivation for sharing.
32. This reprint appeared on March 2, 1852, and was the first to include this particular editorial endorsement, although this language was picked up by many subsequent reprintings of the listicle. The piece in question, "Rules for Home Education," was reprinted in at least 81 newspapers between April 12, 1851 and April 7, 1897.
33. The piece appeared in the *Plymouth Pilot* on May 28, 1851.
34. This piece appeared in the *Democrat and Sentinel* on October 17, 1855, and was reprinted in at least 51 newspapers between January 28, 1853 and January 1, 1864.
35. Franco Moretti, *The Bourgeois: Between History and Literature* (Brooklyn, NY: Verso, 2013), 39.
36. This snippet appears first in our reprinting data in the *Edgefield Advertiser* of July 6, 1842, and was reprinted in at least 81 newspapers. The summary introduction quoted here appears in the reprinting of the *Burlington Free Press*, August 20, 1852.
37. *Going Viral* 72, 94.
38. "Virality 2.0" 543.
39. Our first experiments focused on the pre–Civil War period, which is why we have more extensive network models of it. We are currently developing network models on more recent data sets that span to the end of the nineteenth century.
40. In these graphs, which you can view online, the nodes (the circles) represent individual newspapers, while the edges (the lines between circles) represent texts shared by the newspapers they connect. The edges are weighted based on how many texts two papers share. At http:// networks.

viraltexts.org/1836to1860-filter/index.html you can find a less cluttered version of this 1836–60 graph, in which I filtered the edges so that only relatively strong connections appear. Thus even a very thin line represents several hundred shared texts between publications, while a thick line represents thousands of shared texts. In these graphs, the nodes are larger or smaller based on their "degree," which is a measure of how connected they are to other nodes—in this case, of how many texts they print that other newspapers also print. The curve of the lines attempts to represent the direction of connections—in this case determined by time of reprintings—though that measure is imprecise at best, as we cannot know simply from the days printed whether one newspaper actually got a given piece from a particular other newspaper. To create these models, I used Gephi, which is "an interactive visualization and exploration platform for all kinds of networks and complex systems, dynamic and hierarchical graphs" (https://gephi.github.io/). Gephi is typically used to visualize modern social-media data; hence this application of it to historical text exchange is an active exploration of how and to what extent the modern viral metaphor holds for such a study.

41. "The Inquiry" was also reprinted at least 33 times between November 27, 1840 and March 31, 1860, while "Weights and Measures" was reprinted at least 33 times between October 23, 1851 and December 5, 1860.
42. Bob Nicholson, "'You Kick the Bucket; We Do the Rest!': Jokes and the Culture of Reprinting in the Transatlantic Press," *Journal of Victorian Culture* (2012).
43. Franco Moretti, *Graphs, Maps, Trees: Abstract Models for Literary History* (Brooklyn, NY: Verso, 2005), 53.

Chapter 3

Networking Feminist Literary History: Recovering Eliza Meteyard's Web

Susan Brown

N.B. Additional images associated with this chapter are housed in the digital annex at www.virtualvictorians.org.

This essay addresses how the Internet works for scholars of Victorian literature and literary history, how we work on the Net, how the concept of networks is affecting our engagements with literary history, and how we can make the Net more effective for scholarly purposes. It channels this inquiry through a consideration of Victorian writer Eliza Meteyard, whose case demonstrates how the knowledge networks on which we increasingly draw for our research have profound implications for how that research is shaped.

Meteyard, who wrote under the name of "Silverpen," represents in many ways the opportunities that the nineteenth century presented to a woman of the Anglican professional classes. Although she felt inadequately educated, she lived independently in London on the proceeds of her pen from the early 1840s until her death in 1879. She wrote, not as a leisured lady, but as a professional author. She wrote for money because she needed it to live. She wrote a lot, both nonfiction and fiction, and belonged to the first generation of women to make a living through journalism. In other words, she represents a significant, though not a prominent, figure in Victorian literary history.[1]

Meteyard experienced the conflict between pursuing a writing career and being a woman that recurs throughout the fiction, poetry, letters, and autobiographies of many women of the nineteenth century: she refused marriage and remained single all her life. The heroine in her semiautobiographical novel *Struggles for Fame* also declines a marriage proposal—from a lord, of course—with the assertion that "the woman who wishes to excel in literature must be alone from the cradle to the grave."[2] Yet Meteyard was hardly alone. Her career as a professional writer was only possible because of her position within a complex network of connections that sustained her in this new profession for women, a set of linkages that she actively developed. Meteyard wrote in 1857 of "a spider's web of work, which necessity of ways and means compels me to do."[3] The locution evokes a sense of being both the spider who weaves and the one who is compelled, caught in that same web.

Meteyard's ambivalent image of a web of work is my starting point for exploring how approaches to literary history, and with it humanities scholarship more generally, are being shaped by the Internet and more specifically by the World Wide Web.[4] I will address how feminist literary history is changing and how digital humanities work can advance the analysis of gender and other forms of difference—weaving together the web of work Meteyard invokes to characterize her conditions of possibility as a writer, the work on the World Wide Web that is increasingly part of scholarly life, and the web of words, standards, and technological practices that supports and increasingly defines that scholarly activity.

Narrative literary history was undermined in the late twentieth century by a combination of feminist and other critiques of its conservative and canonizing tendencies, on the one hand, and by the suspicion of narrative that emerged from high theory, on the other hand. Except for focused monographs that investigate some aspect of literary history—recuperating a forgotten writer or filling in a missing context, for instance—it has dwindled into essay-based "companions" to various literary periods and genres and occasional pallid attempts at large-scale narrative history in which, as one reviewer puts it, "narrative gets shouted down by the encyclopedic."[5] Its encyclopedic tendencies perhaps make the Web an unpropitious medium for literary history, particularly given the charge that categories of difference have been notably sidelined within digital humanities.[6] This paper investigates the challenges and potential of Web-based literary-historical inquiry and suggests how we might improve its prospects for doing justice to marginalized subjects, such as Meteyard.

Meteyard is hardly an eminent Victorian. During the composition of this essay, the main Google search engine reported between 81,200 and 135,000 results on her, only 5 or 6 percent of the 1 to 14 million documents it purports to find for Charlotte Brontë. The sparser results in Google Scholar (see annex) suggest that Meteyard's only interest lies in her biography of Josiah Wedgwood, the English pottery manufacturer. These results limited to "scholarly literature"[7] present a narrower profile of Meteyard, with the first nine on the initial page linking to citations or full texts of her *Life of Josiah Wedgwood*, *A Group of Englishmen*, *The Wedgwood Handbook*, and *Wedgwood and His Works*. Only the tenth and last on this first page (or on the second page of results, depending on the day) lists the cryptic "Silverpen— E Meteyard, Closing Movement, Milliner and Dressmakers," hinting at the aspect of Meteyard that is of current scholarly interest. From there, citations link her to the topics of the "fallen" (or sexually compromised) woman, emigration, and female redundancy that have been carefully unpacked by feminist scholars in the past several decades. But you have to dig for these connections.

I focus on Google not only because it owns the most influential search engine in the world,[8] but also these results are just one component of an interlinked suite of networked resources that includes vast quantities of humanities data. It is noteworthy that Google Scholar is out of date. This confirms claims that the digital turn has fundamentally altered the production, dissemination, and authority structures of knowledge: with the advent of Google's PageRank algorithm, links effectively became votes for relevance and value, allowing information to sidestep editorial practices that are standard in both academia and print media generally.[9] Ironically, the traditional criteria of provenance and authority used by Google Scholar seem less effective for scholarly purposes than the basic PageRank algorithm, whose generic search results point to Meteyard's significance drawing on a range of online resources. At one end of the spectrum is Wikipedia, and at the other, the *Oxford Dictionary of National Biography* (*ODNB*), descended from the printed *Dictionary of National Biography*, whose inaugural editor, Leslie Stephen, was the father of Virginia Woolf. Woolf in turn inspired, through her mock literary-historical biography *Orlando*, the Orlando Project—which differs from the *ODNB* in being produced specifically to exploit the potential for digital literary history and which sits somewhere between these two extremes in terms of its mode of production.

Thanks to Meteyard's unusual name, false hits are rare in the first ten pages of Google results. Results include a short biography from the Darwin Correspondence Project, stale information from the print *DNB*, a digitized letter to Leigh Hunt, and links to digitized publications for free or for sale. Summarizing the first few results by resource type, we see an interesting distribution (Table 3.1).

Table 3.1 Summary of Google Search results

	Born-Digital	Surrogate
Academic	*Oxford Dictionary of National Biography* (hybrid surrogate) *Orlando: Women's Writing in the British Isles from the Beginnings to the Present* **Online Books Page (UPenn)**	**The Darwin Correspondence Project** **Leigh Hunt letter**
Nonacademic	**Wikipedia** **Wikisource**	**Internet Archive** **Spectator Archive** Google Books *Amazon, Bookfinder, etc.*

Key: Bold = open resources; Plain = blended resources; Italics = restricted resources

Google offers a much greater wealth of material, particularly in the bottom right quadrant, than does Google Scholar. The Internet Archive leads to 36 monographs by Meteyard, some duplicates but all free, in formats including those intended for the print disabled. Google Books is harder to count, advertising an unlikely 27,500 results, but it certainly provides dozens of Meteyard texts along with a sense of where she stands in current scholarship.[10] At the time of writing, 15 of 20 monographs in the *Cambridge Bibliography of English Literature* were available online for free.[11] Searching for these also led to the discovery of a shorter text, "A Winter's Tears," from the 1861 *The Ladies' Companion and Monthly Magazine*, likely one of many that the *Bibliography* omits.[12]

These resources are nothing short of staggering, especially considering Meteyard's obscurity. Scholars used to have to travel great distances to read such books. I reflect on the digital surrogate of *Dora and Her Papa*, complete with British Library shelf mark, with amazement.[13] Surveying its engraved frontispiece on my home computer screen, I recall frantic reading on research trips, attempts to absorb as much as possible from such books, ordering stacks and stacks of them to storm through as quickly I could, gambling on which to photocopy at steep rates and which to transcribe in part. Such surrogates can't replace physical objects,[14] but still, the Web has brought home much of the print archive, and a sliver of the unpublished archive, in ways that make historical scholarship more accessible than ever. These results are also quite typical. Analyzing the fat author files compiled early in the Orlando Project, we find that most printed texts the project obtained through interlibrary loan or visits to copyright libraries are now online.

The mechanics of literary research have thus been transformed by the Web. To summarize the case for Meteyard: a large proportion of texts are accessible, as is a trail of a few short biographies, a recent article or two, and passing references in a number of other scholarly works. When we consider how we actually use these materials, however, the Web seems to have changed less about how we actually undertake literary history. We can read the books. We can piece together writers' lives from disparate sources, as did the Orlando Project. "History is the essence of innumerable biographies,"[15] according to Thomas Carlyle. The project saw biography as a lens on the material, ideological, and literary factors shaping writers' lives and careers, the codes that, in the words of Janet Todd, "intervene between subjectivity and history and help to fashion both."[16] But a conventional biography still leaves the subject at least formally isolated. Even in collective biography—which, as Alison Booth argues, foregrounds parallels, links, and contrasts—the subjects are formally as separated and insulated as Eliza Meteyard represented the woman writer (and implicitly herself) to be.[17]

Indeed, as Margaret Ezell shows, getting beyond the individual life or text constitutes a significant challenge to literary history.[18] This is the problem foregrounded by Franco Moretti in his contentious advocacy of "distant" reading, based on the fact that we can read only a very small percentage of the books ever published and thus cannot base literary history on a comprehensive knowledge of its object of inquiry.[19] Yet neither Moretti's work nor Matthew L. Jockers's important advances in mining and analyzing large text collections tackle the complex networks of social, intellectual, and literary relations, and their imbrication with material historical conditions, that lie at the heart of literary history—although some projects devoted to understanding networks per se are underway. [20] This slow start in applying social-network tools to literary history may be due to the fact that, as Laura Mandell demonstrates, reading (let alone producing) graphs in order to hypothesize network effects requires considerable expertise and care.[21]

No one is "alone from the cradle to the grave," but their connections are rather elusive. The Web's "links" form a network only in the sense of "a system of interconnected computers":[22] they are like a starburst, with the request emanating out from my computer to Google and then to all those separate files out there on separate servers, which then send files back to my computer. The only link among them is Google's index and the text string "Eliza Meteyard" within it. This helps me to access those distributed resources but doesn't weave them into a meaningful pattern. To do that one must rely on reading, which doesn't scale, as Gregory Crane points out.[23] The task of literary history is to perceive and make sense of the connections lurking in those disparate documents, those little tidbits of information about Meteyard as interconnected with other people, movements, texts, political (or ideological or technological or material) changes, and other relevant phenomena—which is to say, with history. The Net does not help much here.

The Orlando Project adopted traditional methods of analog literary history: weaving together strands from many and disparate sources to create larger patterns. Although the emergent research web sped things up, the literary historical method was not substantially changed by our use of computers. It was, however, changed by how we wove that web into our writing. So to start tracing Meteyard's web, I turn to the published textbase *Orlando: Women's Writing in the British Isles from the Beginnings to the Present.*

Eliza Meteyard's entry in *Orlando* indicates a biographical account of a woman writer's life belonging to the genre that Alan Liu characterizes as "*pétits recits*," little narratives that are not linked by larger claims or arguments.[24] Its underlying form, indeed, is not even a single document (see annex), a fact signaled by the division of the entry into tabbed screens labeled "Overview," "Writing," "Life," "Life & Writing," "Timeline," "Links," "Links Excerpts," and "Works By." The "Overview" screen is composed of a brief summary of Meteyard's significance, several chronological "Milestones," which will also appear embedded in the narrative "Life" and "Writing" screens as well as in the "Timeline" of all chronological items associated with Meteyard, and hyperlinked headings leading to the sections of the "Life" and "Writing" narratives. Symptomatic of the database, which, Ed Folsom claims, is "the genre of the twenty-first century" whose "development may turn out to be the most significant effect computer culture will have on the literary world,"[25] this "entry" is dynamically generated from a number of files and database elements. Media theorist Lev Manovich says that "database and narrative are natural enemies."[26] However, as Folsom argues, when a narrative sits on top of a database, there is a tension between the two that is productive of new kinds of knowledge that neither would offer alone.

Orlando sustains this tension between database and narrative in several ways. It does this, first, in highly granular citations interspersed in the prose and, second, in links that offset defined narrative trajectories with less predictable ones that might qualify as "ergodic" in requiring an active choice of paths.[27] Every person, place, organization, and title in *Orlando* is hyperlinked to a screen that gathers links to all other occurrences of those entities. Sometimes, as with Meteyard's publication vehicle *Tait's Edinburgh Magazine*, the "Links" screen for that entity points readers toward other narratives or composite sequences generated from snippets of multiple narratives: in this case 46 links to passages in 15 different entries. Third, *Orlando* weaves together database and narrative through semantic encoding or tagging. Here is an excerpt from the encoded text:

> <CHRONSTRUCT CHRONCOLUMN=BRITISHWOMENWRITERS RELEVANCE=PERIOD ID=w--meteel--0--CHRONSTRUCT--1 > <DATE VALUE=1846-- CERTAINTY=C > Around 1846 </DATE> <CHRONPROSE > <PPERIODICALPUBLICATION > <PAUTHORSHIP AUTHORNAMETYPE=PSEUDONYMOUS > <NAME STANDARD=Meteyard, Eliza > EM </NAME> gained her pen name of <SOCALLED > Silverpen </SOCALLED > when <NAME STANDARD=Jerrold, Douglas William > Douglas Jerrold </NAME> appended it to a leading article she contributed to his <TITLE TITLE-TYPE= JOURNAL REG=Douglas Jerrold's Weekly Newspaper> Weekly Newspaper </TITLE>. <BIBCITS > <BIBCIT PLACEHOLDER= DNB DBREF=1759> </BIBCIT> </BIBCITS> </CHRONSTRUCT>

This encoding underlies the readable text and provides the means of databasing it. The encoding points to context, regardless of the actual readable words, much as a printed book's index does, providing a structured navigational lens that complements the continuous text, drawing discontinuous portions of the text under the same conceptual banner. So, for instance, *Orlando* labels Meteyard's pen name "Silverpen" as a pseudonym and also tags discussions of pseudonymous publication within her entry. The encoding is grouped with similar discussions, of which there are 948 for women who lived during the Victorian period. A few (simplified and reduced; see annex) excerpts from this set read as follows:

> **Charlotte Brontë** 1816-1855
> 29 December 1836 CB solicited Poet Laureate Robert Southey's opinion on some poems; he advised her to pursue her "proper duties," because "Literature cannot be the business of a woman's life, and it ought not to be." Bibliographic Citation link.
> Emily, Anne, and CB published a collection, *Poems*, under the pseudonyms Currer, Ellis, and Acton Bell.
> 16 October 1847 CB's novel *Jane Eyre: An Autobiography* was published in three volumes as "edited by Currer Bell."
> Rigby also responded to the widespread speculation that 'Currer Bell' was both a woman and a governess with the view that the book

she deplores for an inexcusable "coarseness of language and laxity of tone" exhibits minutiae with regard to women's dress "which at once acquit the feminine hand." She adds that if the author was a woman, she must have "long forfeited the society of her own sex." CB wrote a sarcastic response in which she advised the author to try the "unpitied suffering" and "uncheered solitude" of governessing for a couple of years. Smith, Elder refused to use this as a preface for *Shirley*.

7 July 1848 CB travelled to London with her sister Anne to refute the claim that Currer, Ellis, and Acton Bell were a single author.

Eliza Meteyard 1816 - 1879

Around 1846 EM gained her pen name of 'Silverpen' when Douglas Jerrold appended it to a leading article she contributed to his *Weekly Newspaper*.

By 1846 she was also using her own name in publications, billing herself as the author of *Struggles for Fame*.

Situating Meteyard within such database results recontextualizes "Silverpen" in relation to different types of pseudonyms, as well as to other aspects of authorship ranging from the often overlapping use of anonymity to collaborative writing.[28] Such juxtapositions allow us to explore literary-historical claims, such as Dorothy Mermin's assertion that "female novelists' use of male pseudonyms . . . characterize[s] the period from the 1840s to 1880."[29] This assertion, Mermin makes clear, resides within a broader context in which anonymity was normalized in certain areas of publishing, such as periodicals, and in which different kinds of pseudonyms did different cultural work. These sorts of complexities are evident in the *Orlando* results. For instance, although Charlotte Brontë's pseudonym is associated with *Jane Eyre*, *Orlando* links it initially with the publication of a book of poems, and Meteyard's pseudonym is related to publishing first as a journalist. We can also deduce that an obvious fabrication like "Silverpen" differs from names that resemble those of actual people, whether gender-ambiguous ones like "Currer Bell" or gender-precise ones like "Diana Butler."

As a narrative argument, Mermin's book asserts these complexities while offering evidence that is necessarily and, Stanley Fish would argue,[30] fruitfully selective. Yet I for one regret that we can't access the larger body of Mermin's evidence. This is not unlike most computational scholarship to date, though: Gaye Tuchman and Nina Fortin's early pioneering study, for instance, investigated the same question and concluded that in the 1860s and 1870s men were more likely to use pseudonyms to masquerade as women than the reverse.[31] However, *Edging Women Out* gives no more access to its evidence than does Mermin or for that matter Moretti. The separation of the database from the narrative thus makes the conclusions reached by both conventional and digital methods less convincing. I am not arguing for some kind of new "scientism" of digital literary history. Quite the contrary: the application of computational processes to humanities unavoidably brings the burden of interpretation, requiring the ability to examine the underlying "data" or evidence that has produced the patterns we interpret.[32] But very few experiments using immense corpora are currently able to link their claims to the evidence in such a way that the reader, as opposed to the scholar, can so move.[33] Such links are key, as Folsom argues, to troubling the narratives emerging from databases. Matthew Jockers closes his foray into literary history by remarking that "these network data are rich—too rich, in fact, to take much further in these pages because they demand that we follow every macroscale observation with a full-circle return to careful, sustained, close reading."[34] But it remains the case that we have less access to the evidence for his arguments than we would if they were based on a series of traditional close readings. Equally critical is the ability to evaluate the corpora on which results are based, since as Katherine Harris, Amy Earhart, Keith Lawson, and Marjorie Stone have shown, many corpora are skewed against marginalized groups.[35] Although pseudonyms present a problem in contextual analysis, not one in textual analysis, the principle of connecting the macro and the micro is equally crucial when we turn to methods such as social-network analysis.[36]

METEYARD'S WEB OF WORK

Turning to the rich contextual evidence now available online for Eliza Meteyard, one thing we can deduce is that she did not seek, as did George Eliot and Charlotte Brontë, to use her pseudonym Silverpen to disguise either her gender or her identity. Within the literary world, at least, she was at pains to reveal them. Thus we find, in the digital facsimile of her letter to Leigh Hunt (see annex), evidence of what we would now call active networking in the service of her literary career. The letter reads:

> 59 Lamb's [?Conduit] Street
> June 26. 1848
> Sir
> Some two years and a half ago since, I wrote to you about a review in the Examiner, and you were kind enough to return me a cordial

> answer - I have that answer now. Since then I have [?merged] into the "Silverpen" of Douglas Jerrold's Newspaper - and by him, best and first of friends, so named.
> I do not write to inform you of this fact, but that, sending with this two numbers of Howitt's Journal, it is to apologize for the, with me, unusual liberty of so troubling any one. But within these numbers a little story his, which having
> [page break]
> some foundation in truth I should like you to read, not for work of mine therein, not a line of it, but that it will be fresh evidence again to you, a true poet, that there is something glorious in this human heart of ours! And when we hold a need [?aft] to be laughed at as visionary, it is pleasant for testimony to rise up again, and again, no matter how often, in demonstration of what we know to be an eternal truth. The story of Lord Burleigh has always been a pleasant one to me.
> As a child, I had knelt upon those old altar stones in Hadnal[l]
> [page break]
> Church in Shropshire, where years before knelt that true Lord and that sweet nature's Lady.
> Pardon this impression you will, if I judge aright the portrait of you, I saw some days since in my friend Margaret Gillies studio.
> With highest respect, believe me,
> Sir
> Your's faithfully
> Eliza Meteyard[37]

She reminds Hunt, an influential critic and publisher (and the original of Harold Skimpole in *Bleak House*), in only 252 words, of their previous correspondence, her emergence into professional authorship, her pseudonym, and her patronage by the publisher Douglas Jerrold. She sends a writing sample, simultaneously performs humility, critical acumen, and womanly affect, and as a coup de grâce invokes her connection to a well-known literary and social justice network by complimenting a portrait of Hunt himself.

Meteyard goes a bit over 140 characters, but one couldn't do a better job of self-promotion on Twitter. Here we see her firming up what social-network analysts call *weak links* and boasting of strong ones. If the allusion to the journalist and playwright Jerrold, author of the wildly popular play *Black-Eyed Susan* and friend of Dickens, wasn't sufficient, the allusion to Hunt's portrait by Margaret Gillies would ensure he took note. Gillies was a prolific and successful Victorian artist, a feminist, and a member of highly influential overlapping aesthetic and social circles. She links Meteyard to utilitarianism, radical and reform politics, radical Unitarianism, and feminist and protolesbian communities. Gillies's common-law partner Thomas Southwood Smith—a friend of Hunt's—pursued social reform through publishing, as a founder of the Society for the Promotion of Useful Knowledge and as a commissioner of Parliament into the state of children's employment in Britain; just so, Gillies broke new ground for women as a prolific artist who exhibited at the Royal Academy and internationally. In the 1840s she moved from an early career in miniatures to larger, more ambitious canvases, including of subjects like Hunt. She also apparently contributed to the illustrations of the Reports of the Children's Employment Commission that launched a sex scandal regarding the working conditions for women and girls in the mines.[38] By 1861 *Art Journal* praised Gillies as unparalleled in her "quasi-classic kind of art."[39] Additionally, she rejected the segregation of private clubs as a member of the Council of the Whittington Club, a radical experiment in which members were welcome regardless of class or sex, and to which Meteyard belonged along with Mary Howitt, Mary Cowden Clarke, Harriet Martineau, Eliza Cook, and Camilla Toulmin.[40]

Gillies, just one node in Meteyard's network, points to a myriad of affinities and values—not least radical politics, belief in art for social change, and an emancipated perspective that countenanced a married man and a single woman living together openly. Weaving such connections from the tendrils of this loosely connected web of online material led me to some new insights about Meteyard, as did recent print publications. In both cases, such insights rely on prior knowledge of context and the ability thereby to infer connections among scattered and siloed materials.

One of the big questions of computational analysis is how to arrive at that which we do not already know, to get machines to help on the path to inference.[41] The kind of semantic markup embedded in *Orlando* materials provides one strategy for moving in that direction. This markup allows us to explore webs of relationships that have meaning built into their links through the semantic tags.[42] The network visualization tool OVis creates a standard network graph that visualizes people or nodes as boxes and the links between them as lines, evoking a better sense of the extent and complexity of Meteyard's networks than the other resource types and representations of links considered thus far.

OVis can display links that are defined by broad connections among people—for instance, Meteyard is connected to radical Unitarianism as represented by W. J. Fox, Margaret Gillies's partner Smith, and feminists such as Barbara Leigh Smith. Even drawing only on

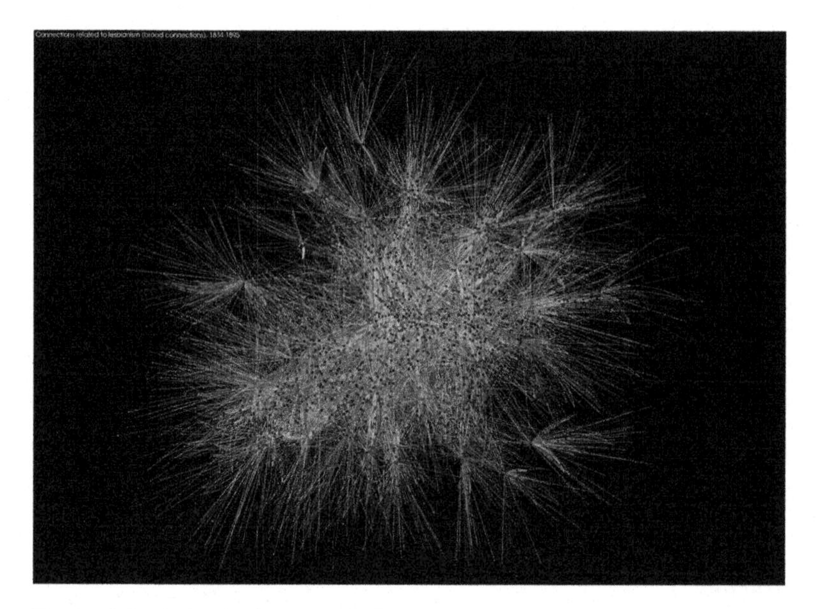

Figure 3.1 OVis visualization of all connections between people connected to lesbianism.

the material in *Orlando* produces a large Unitarian network within OVis (see annex). The graph shows a densely connected network of thousands of lines representing relationships connecting thousands of individuals. Without more tools than this interface affords for exploring the variances in proportions of nodes, link types, or path lengths, it is hard to make sense of the evident differences between this network, for instance, and the similarly large one showing the thousands of relationships related to the protolesbian subculture represented by Eliza Cook, Matilda Hays, and Charlotte Cushman.

More useful are restricted sets of relations, such as those revealed by limiting the graph to links among Meteyard, other writers, and Unitarianism defined more narrowly by co-occurrence within the same chunk of text as the interpersonal links. This graph (see annex) shows connections to six other authors: one each to Harriet Martineau and Julia Wedgwood, two to Eliza Cook, three each to Elizabeth Gaskell and Matilda Hays, and nine to Mary Howitt. These relationships map, through representative women, Meteyard's preoccupations with radical politics and social justice, with emergent forms of domestic realism, with woman-centered communities, and with feminism—as well as her links to the Darwin and Wedgwood families. The graph

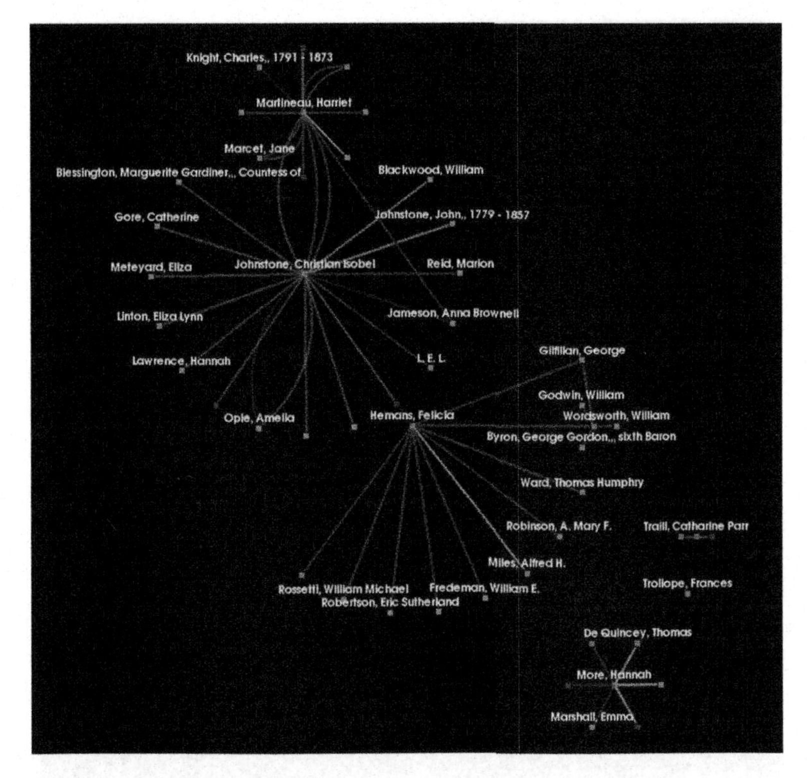

Figure 3.2 OVis visualization of connections to *Tait's Edinburgh Magazine*.

supports Kathryn Gleadle's claim that Unitarianism laid the foundation for midcentury feminism through "essential ideologies and personnel networks."[43]

Because the OVis tool was created to leverage the power of *Orlando*'s semantic markup, we can read these more focused graphs for the kinds of relationships involved: the links are colored to indicate the contexts of the connections. Here it is useful to compare two early feminist venues in which Meteyard published.

The populist *Tait's Edinburgh Magazine*, with Christian Isobel Johnstone as editor from 1834 to 1846, did much, as Alexis Easley has established, to advance women's journalism.[44] The OVis graph (see annex for color) of connections to *Tait's* is quite femalecentric. It displays a diffuse network with few interconnections, and an almost entirely literary one, with the blues through pinks representing links related to literary production and reception. The two green lines trace Johnstone's

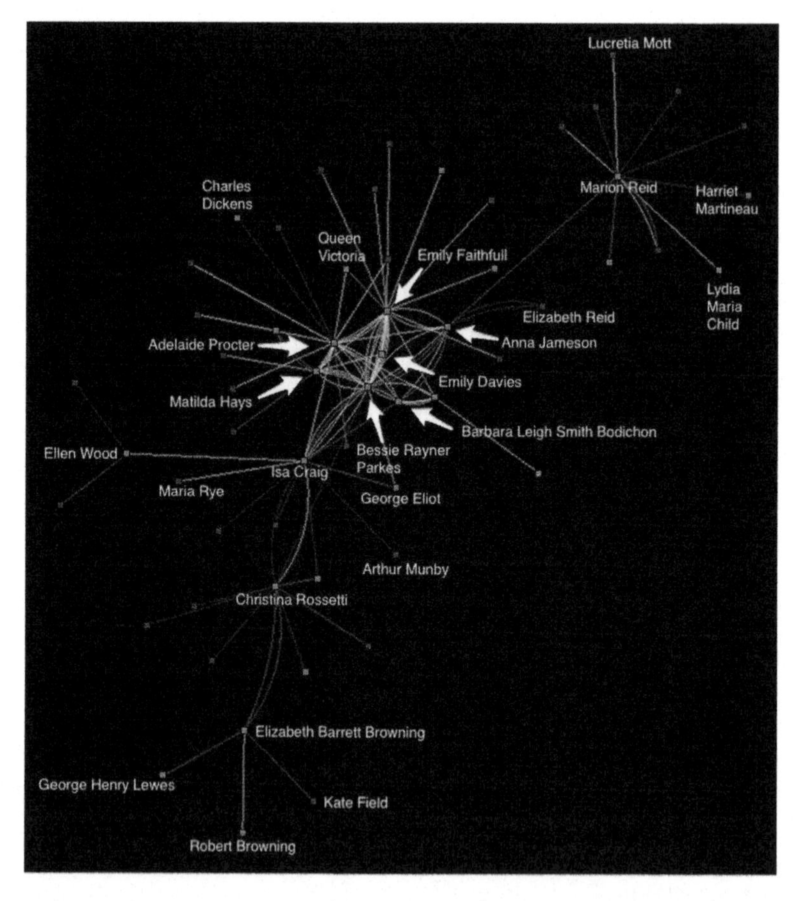

Figure 3.3 OVis visualization of network connected to the *English Woman's Journal.*

relationships to her husband and collaborator, John Johnstone, and to William Blackwood, who co-owned another journal she edited.

If we compare the *Tait's* graph to the graph for the groundbreaking protofeminist *English Woman's Journal,* we see a wider range of link types, with lots of green (see annex for color) for social relations and many more interconnections between people, in what is a busier and tighter network.

These graphs are in no way comprehensive or definitive representations, although the tool at least makes it possible to read the text from *Orlando* on which the links are based. What they tell is not entirely new: as indicated, the implications of some of these connections are already noted by scholars. Yet these visualizations, limited though they are, allow us to glimpse the power of such network graphs to bring together large quantities of granular information from disparate sources in a way that facilitates contextual exploration and analysis. They overlap and intersect, leading us to ask several questions: Which links are strong and persistent, and which weak and ephemeral? What people link disparate groups to each other? How do these graphs compare with views of other groups in this period, such as those in the poetess movement or those associated with literary annuals? They suggest the potential that emerges when links themselves become meaningful, legible, interpretable, selectable, and excludable, in contrast to the Web links we have now that are largely devoid of content.

SEMANTIC WEBS

Meaningful links are the essential feature of the semantic web, a set of technologies for producing more interoperable and machine-processable data for the Web that has been promoted by Tim Berners-Lee and others since 2001.[45] One of the more promising developments in digital humanities is that scholars are beginning to create and adapt resources for the semantic web. The single largest and most versatile open resource for nineteenth-century studies available today, NINES, uses lightweight semantic web technology to give combined access to almost a million items from both open and licensed collections.[46] Indeed, searching in NINES adds new strands to the web of Meteyard's connections, Dickens and Gladstone among them. However, humanities scholars have only tentatively ventured into the semantic web, and there is much at stake in how we do. Most humanities projects are still siloed, experimenting with contained sets of semantic web data and fairly basic vocabularies rather than engaging with the thorny questions of how to link their knowledge domains to other areas of the semantic web. Without such engagement, the semantic web's promise to bring together disparate sources and allow the inference of new information based on their combination is lost, as is its ability to reflect humanities epistemologies.

As the semantic web grows, its ability to handle categories of identity and of difference will become crucial. Tara McPherson argues compellingly that there are profound historical reasons for the resistance of digital humanities to race theory and, indeed, to the theoretical and ethical imperatives that have shaped literary studies since the 1980s. She links the development of the modular programming philosophy that underlies the UNIX operating system to the post-war management of racial relations through "lenticular" logics that manage what kinds of relations or connections are perceptible at one time; this connection is made via an analogy with the lens technology that conjoins two images in a single material object but segregates one vision from one another, as in postcards where one sees different images at different angles.[47] It is reasonable, I would argue, to extend McPherson's argument regarding race to gender on the basis of connections between the civil rights movement and the feminist movement. Doing so helps to explain why the "feminist" side of digital humanities has to date resided almost entirely in content, or in the "soft side" of the field, in its margins rather than its center. This argument goes beyond the notion of the interrelation of textual form and content explored in recent work on anthologies or serialization, for example, to suggest why modes of technological production seem insulated from political objectives. Jacqueline Wernimont has argued that we have yet to parse this relationship, and Alan Liu and Martha Nell Smith have both suggested the limitations of modular interfaces in similar contexts.[48]

This is not to say that we lack ideas about how feminist engagement with the epistemologies that shape our technologies might look. For Donna Haraway, webs are the crucial image. She argues that partial or situated knowledges can lead to "better accounts of the world, that is, 'science.'"[49] In her view,

> webs can have the property of being systematic, even of being centrally structured global systems with deep filaments and tenacious tendrils into time, space, and consciousness, which are the dimensions of world history. Feminist accountability requires a knowledge tuned to resonance, not to dichotomy. Gender is a field of structured and structuring difference, in which the tones of extreme localization, of the intimately personal and individualized body, vibrate in the same field with global high-tension emissions. Feminist embodiment, then, is not about fixed location in a reified body, female or otherwise, but about nodes in fields, inflections in orientations, and responsibility for difference in material-semiotic fields of meaning.[50]

This kind of knowledge representation aims to challenge the stance of objectivity through a focus on method, but it may also threaten its own epistemological roots, since, McPherson argues, a movement like feminism, which takes gender as a category of analysis, emerges from the same logic of modularity as contemporary programming:

> We might also understand the emergence of identity politics in the 1960s as a kind of social and political embrace of modularity and

> encapsulation, a mode of partitioning that turned away from the broader forms of alliance-based and globally inflected political practice that characterized both labor politics and antiracist organizing in the 1930s and 40s.[51]

Haraway's refusal to fix the grounds of feminism may be read as a refusal of modularity in favor of a field of structured and structuring difference that informs inquiry into both the materiality and the semiotics of culture. A practical application of this position might involve trying to build inflection, orientation, flexibility, and difference into the emergent structures of the semantic web. This is of paramount importance, since those structures will not only determine how the connections among cultural materials are made, but will also shape the lenses through which we view them—within visualizations, for instance.

The semantic web will be governed by the vocabularies that define its components and the tools built to use them. Definitions of "things"—concepts, entities, named phenomena—and the relationships or links among them will imbue those things with digital reality, determining what it is possible to perceive and represent, and will place them within conceptual hierarchies and structures. Vocabularies will establish the logic by which machines relate things, such as Meteyard and Unitarianism or lesbianism. Right now, standard semantic web vocabularies are impoverished with respect to describing cultural materials, their histories, and their relationships with the rest of the world,[52] despite the fact that the amount of available linked data is growing exponentially and is manifesting in embryonic ways in major Web resources.

One such manifestation is the Knowledge Graph, introduced by Google in May 2013 as part of their proclaimed shift "from an 'information engine' to a 'knowledge engine.'" Their explanation of this enhanced search of "things, not strings," stresses the power of the "intelligence between" things.[53] The Knowledge Graph creates a mashup of information by bringing together Google's own resources and algorithms with a number of existing semantic web resources to create a graph consisting of "more than 500 million objects, as well as more than 3.5 billion facts about and relationships between these different objects."[54] This information comes from sources such as DBpedia (a semantic web resource derived from Wikipedia), Freebase (a knowledge base owned by Google), and other linked data sources. Knowledge Graph results are currently displayed in a visually distinct box, on the top right of Google search results screens in the form of text and thumbnail images.

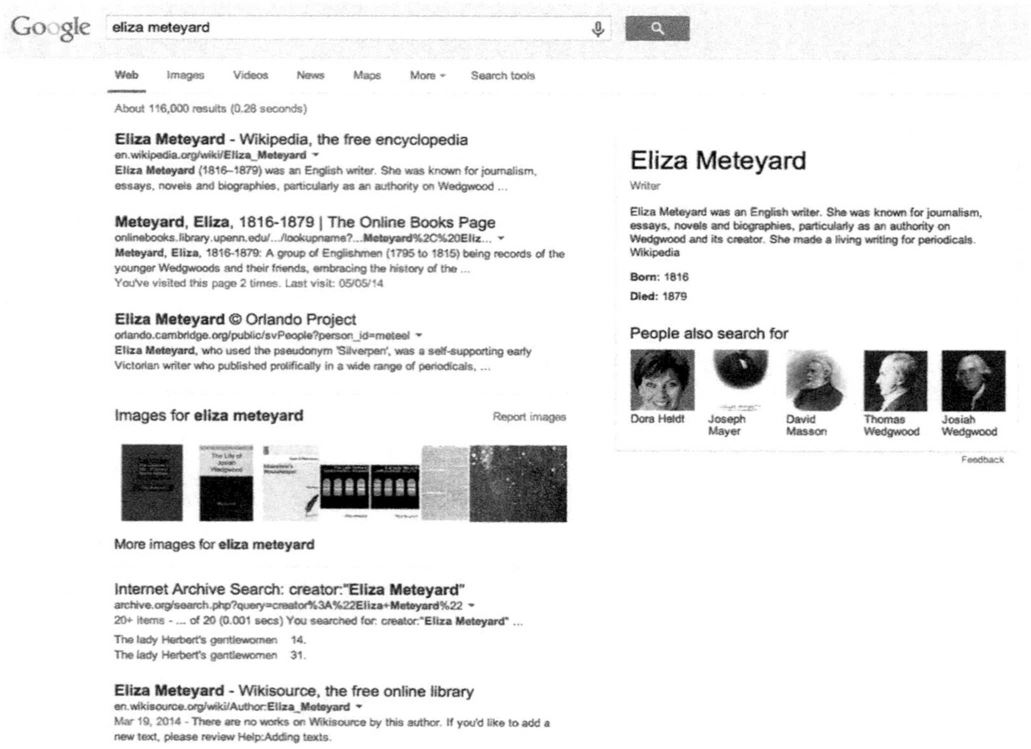

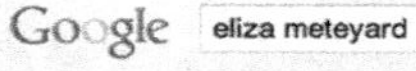
eliza meteyard
Web Images Videos News Maps More Search tools
About 116,000 results (0.28 seconds)
Eliza Meteyard - Wikipedia, the free encyclopedia
en.wikipedia.org/wiki/Eliza_Meteyard
Eliza Meteyard (1816–1879) was an English writer. She was known for journalism, essays, novels and biographies, particularly as an authority on Wedgwood ...
Meteyard, Eliza, 1816-1879 | The Online Books Page
onlinebooks.library.upenn.edu/.../lookupname?...Meteyard%2C%20Eliz...
Meteyard, Eliza, 1816-1879: A group of Englishmen (1795 to 1815) being records of the younger Wedgwoods and their friends, embracing the history of the ...
You've visited this page 2 times. Last visit: 05/05/14
Eliza Meteyard © Orlando Project
orlando.cambridge.org/public/svPeople?person_id=meteel
Eliza Meteyard, who used the pseudonym 'Silverpen', was a self-supporting early Victorian writer who published prolifically in a wide range of periodicals, ...
Images for eliza meteyard
Report images
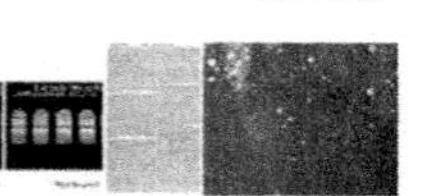
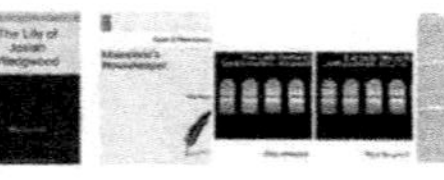
More images for eliza meteyard
Internet Archive Search: creator:"Eliza Meteyard"
archive.org/search.php?query=creator%3A%22Eliza+Meteyard%22
20+ items - ... of 20 (0.001 secs) You searched for: creator:"Eliza Meteyard" ...
The lady Herbert's gentlewomen 14.
The lady Herbert's gentlewomen 31.
Eliza Meteyard - Wikisource, the free online library
en.wikisource.org/wiki/Author:Eliza_Meteyard
Mar 19, 2014 - There are no works on Wikisource by this author. If you'd like to add a new text, please review Help:Adding texts.
Eliza Meteyard
Writer
Eliza Meteyard was an English writer. She was known for journalism, essays, novels and biographies, particularly as an authority on Wedgwood and its creator. She made a living writing for periodicals. Wikipedia
Born: 1816
Died: 1879
People also search for
Dora Heldt
Joseph Mayer
David Masson
Thomas Wedgwood
Josiah Wedgwood
Feedback

Figure 3.4 Excerpt from Google search results for Eliza Meteyard as of May 2014. Google and the Google logo are registered trademarks of Google Inc., used with permission.

The Knowledge Graph leverages straightforward information well. The facts provided for Meteyard (the label "Writer," a brief description from Wikipedia, and her birth and death dates) or for the more documented Charlotte Brontë are accurate, and the Knowledge Graph provides a handy means of allowing users to disambiguate the results for entities with the same name (for example, the birds called *penguins* versus the Pittsburgh Penguins). A mechanism, later removed, initially allowed users to contribute to disambiguation, for instance by reporting mislabeled images. However, what if the connection between an entity and an image is more complicated than simple mislabeling? Identifying pictures of Katharine Bradley and Emma Cooper as "Michael Field" isn't precisely wrong, but the relationship between them and their alter ego isn't defined as carefully as it should be. Likewise, it is not true, as the graph states, that Michael Field attended Newnham College, though Bradley did—before "Michael Field" even existed.

If we take the Knowledge Graph as the most accessible portion of the semantic web to date, and an indicator of its eventual impact, there is cause for concern when it comes to nuance and complexity. For instance, Google provides a Knowledge Graph for *World War Z* (the zombie movie) but not for World Wars I and II. Entities such as Jesus and Palestine are hugely controversial, as we know from Wikipedia edit wars, and neither has extensive Knowledge Graphs. This may indicate that complex or contested entities, because they are harder to model, will be left out or shortchanged. Ambiguity, controversy, and outliers can be difficult to detect in large graphs, particularly when data is heterogeneous or dirty, and figures such as Field offer boundary cases that illustrate the extent to which some widely used semantic web ontologies fail to reflect complexity.[55] Even the exemplary Virtual International Authority File, derived from national library catalogues, is pushing out linked data that flattens and misrepresents relationships among different contributors to the production of books.[56]

Humanities scholars are experts in diversity, particularities, nuances, and local meaning. So here is a major opportunity: not simply to critique the limitations of the semantic web but to recognize the immense potential offered by meaningful links and to embrace the challenge of figuring out how to build vocabularies representing entities and relationships in ways, to return to Haraway, that are "tuned to resonance, not to dichotomy."[57] One aspect of the Knowledge Graph touted by Google is the autocomplete functionality that can help disambiguate entities as you type. However, this very feature has been identified as one that reifies and perpetuates racism and misogyny.[58] To work against such tendencies, humanities scholars will require tools that allow us to navigate and manipulate the vocabularies and valences of difference productively, in a bid to represent the uneven developments that make up literary history. The complex network of labor relations, friendships, politics, sexuality, spirituality, and textuality surrounding Eliza Meteyard, for instance, constitutes a dynamic web of differences held in tension.

The semantic web is more modular than any other attempt at providing the basis for interoperability and integration of materials on the Web. It may suffer from lenticular logic. Certainly we cannot yet fully evaluate the implications of particular vocabulary or ontological choices. Yet we can draw on extensive humanities work in epistemology and representation as we aim to formalize the kind of situated knowledge advocated by thinkers such as Haraway. The links between narrative and database, form and content, abstracted encoding or metadata and source materials, are key. It is crucial to close the gap between scholarly consumption of resources and their production and ongoing refinement: the semantic web needs the expertise of scholars attuned to respecting differences and also probing the relationships among them. We can do that by adding to the pool of linked data, either by exposing existing data in that format or by creating new resources. As humanities researchers engage with emergent forms of digital work, we can strive to incorporate responsibility for difference into the material-semiotic fields of meaning that increasingly define our research. As did Meteyard, we can actively shape, expand, and develop the networks within which we move, so that we find ourselves enabled even as we are necessarily constrained. Success would mean the ability to enact new literary histories, among other forms of inquiry.

Another pseudonymous Victorian woman writer, herself a node in Meteyard's own network, weighed in on the question of approaches to history. George Eliot, contrasting her method to that of Henry Fielding, famously proclaimed:

> I at least have so much to do in unravelling certain human lots, and seeing how they were woven and interwoven, that all the light I can command must be concentrated on this particular web, and not dispersed over that tempting range of relevancies called the universe.[59]

Yet, by focusing on one particular web, Eliot reflected profoundly on the "relevancies" of the universe. She encapsulates the tensions between the particular and the general, the concentrated and the dispersed, around which these reflections on literary history revolve. In this she offers another meaningful link that brings the courageous, socially committed world that spawned the professional woman writer of the Victorian age into conversation with the scholarly and technological challenges of our own time.

NOTES

1. "Eliza Meteyard," *Orlando: Women's Writing in the British Isles from the Beginnings to the Present*, ed. Susan Brown, Patricia Clements and Isobel Grundy (Cambridge: Cambridge UP, 2006–14), accessed April 28, 2014, http://orlando.cambridge.org. Fred Hunter, "Meteyard, Eliza (1816–1879)," *Oxford Dictionary of National Biography* (Oxford: Oxford UP, 2004), accessed April 28, 2014, doi:10.1093/ref:odnb/18624.
2. Eliza Meteyard, *Struggles For Fame* (London: T. C. Newby, 1845), 367.
3. Ronald W. Lightbown, "Introduction," *The Life of Josiah Wedgwood* (London: Cornmarket Press, 1970): n.p.
4. The two are not the same: the World Wide Web is the global set of documents linked by the TCP/IP and HTTP protocols and associated software, while the Internet is the set of interconnected computing networks that make the Web possible.
5. James Wood, "The Slightest Sardine," review of *The Oxford English Literary History, Vol. XII: 1960–2000: The Last of England?*, by Randall Stevenson, *London Review of Books*, May 20, 2004, accessed May 5, 2014, http://www.lrb.co.uk/v26/n10/james-wood/the-slightest-sardine.
6. Alan Liu, "Where Is Cultural Criticism in the Digital Humanities?" in *Debates in the Digital Humanities*, ed. Matthew K. Gold (Minnesota: University of Minnesota Press, 2012): 490–509. Various, Twitter threads, 2012–14 "#pocodh" and "#transformdh," https://twitter.com/search?q=%23pocodh&src=typd and https://twitter.com/search?q=%23transformdh&src=typd. Postcolonial Digital Humanities, last modified April 2014, accessed May 5, 2014, http://www.dhpoco.org. #TransformDH, last modified May 5, 2014, accessed May 5, 2014, www.transformDH.org.
7. "About Google Scholar," Google, accessed April 28, 2014, http://scholar.google.com/scholar/about.html.
8. "comScore Releases July 2013 U.S. Search Engine Rankings," comScore, Inc., accessed October 23, 2013, http://www.comscore.com/Insights/Press_Releases/2013/8/comScore_Releases_July_2013_U.S._Search_Engine_Rankings. Danny Sullivan, "Google Still World's Most Popular Search Engine by Far, but Share of Unique Searchers Dips Slightly," February 11, 2013, accessed May 5, 2014, http://searchengineland.com/google-worlds-most-popular-search-engine-148089.
9. Bill Cope and Mary Kalantzis, "The Role of the Internet in Changing Knowledge Ecologies," *Arbor* 737 (2009): 521. Lawrence Page, Sergey

Brin, Rajeev Motwani, and Terry Winograd, "The PageRank Citation Ranking: Bringing Order to the Web," *Stanford InfoLab* (1999), accessed May 5, 2014, http://ilpubs.stanford.edu:8090/422/.

10. Google Books offers a mixture of open, preview, and "snippet" views; most of the texts by Meteyard, however, are open. There is significant overlap here with the Internet Archive, since it has uploaded about a million Google books into its "open public library."
11. Joanne Shattock, ed., *The Cambridge Bibliography of English Literature: 1800–1900*, vol. 4 (Cambridge: Cambridge UP, 1999).
12. Silverpen [Eliza Meteyard], "A Winter's Tears," *The Ladies' Companion and Monthly Magazine*, vol. 19 (London: Rogerson and Tuxford): 18–23, accessed February 12, 2015, http://books.google.ca/books?id=KEYFAAAAQAAJ&pg=PP7&lpg=PP7&dq=%22The+Ladies%E2%80%99+Companion+and+Monthly+Magazine%22&source=bl&ots=YFWcbbhbkS&sig=ZnNbmbBLt14e3K7yjPkM6Nz-j_Q&hl=en&sa=X&ei=u4leU8zaFaH62gW_1YG4Bg&ved=0CEAQ6AEwBg#v=onepage&q=Winter's&f=false.
13. Eliza Meteyard, *Dora and Her Papa* (London: George Routledge and Sons, 1869), Google books edition, accessed April 28, 2014, http://books.google.ca/books?id=VMIBAAAAQAAJ&printsec=frontcover&dq=Dora+and+her+papa&hl=en&sa=X&ei=FKxiU73qJdKhyAST8ICoCA&ved=0CC8Q6AEwAA#v=onepage&q=Dora%20and%20her%20papa&f=false.
14. Lisa Gitelman, *Always Already New: Media, History and the Data of Culture* (Cambridge, MA: MIT Press, 2008): 143–48. Marjorie Stone and Keith Lawson, "'One Hot Electric Breath': EBB's Technology Debate with Tennyson, Systemic Digital Lags in Nineteenth-Century Literary Scholarship, and the EBB Archive," *Victorian Review* 38, no. 2 (2012): 101–26. Lynn C. Westney, "Intrinsic Value and the Permanent Record: The Preservation Conundrum," *OCLC Systems & Services* 23, no. 1 (2007): 10.
15. Thomas Carlyle, "Thoughts on History," *Fraser's Magazine for Town and Country* 10 (1830): 414.
16. Janet Todd, *Feminist Literary History: A Defence* (Cambridge: Polity Press / Basil Blackwell, 1988), 137.
17. Alison Booth, *How to Make It as a Woman: Collective Biographical History from Victoria to the Present* (Chicago: University of Chicago Press, 2004).
18. Margaret Ezell, *Writing Women's Literary History* (Baltimore: John Hopkins UP, 1996).
19. Franco Moretti, *Graphs, Maps, Trees: Abstract Models for a Literary History* (London and New York: Verso, 2005).
20. For example, Six Degrees of Francis Bacon is attempting to recreate a social network by mining existing scholarship for relationships. Six Degrees of Francis Bacon, n.d., accessed May 5, 2014, http://sixdegreesoffrancisbacon.com/overview.

21. Laura Mandell, "How to Read a Literary Visualisation: Network Effects in the Lake School of Romantic Poetry," *Digital Studies/Le champ numérique* 3, no. 2 (2013).
22. Oxford English Dictionary, s.v. "network," accessed April 28, 2014, http://www.oed.com.
23. Gregory Crane, "What Do You Do with a Million Books?" *D-Lib Magazine* 12, no. 3 (March 2006), accessed April 28, 2014, doi:10.1045/march2006-crane.
24. Alan Liu, *The Laws of Cool: Knowledge Work and the Culture of Information* (Chicago: University of Chicago Press, 2004).
25. Ed Folsom, "Database as Genre: The Epic Transformation of Archives," *Publications of The Modern Language Association of America* 122 (2007): 1576.
26. Lev Manovich, *The Language of New Media* (Cambridge: MIT P, 2001): 225.
27. Espen J. Aarseth, *Cybertext: Perspectives on Ergodic Literature* (Baltimore: John Hopkins UP, 1997).
28. This encoding within Extensible Markup Language (XML) embeds the intellectual principles of the project. Susan Brown, Patricia Clements, and Isobel Grundy, "Sorting Things In: Feminist Knowledge Representation and Changing Modes of Scholarly Production," *Women's Studies International Forum* 29, no. 3 (2006): 317–25.
29. Dorothy Mermin, *Godiva's Ride: Women of Letters in England, 1830–1880* (Indiana: Indiana University Press, 1993), 48.
30. Stanley Fish, "The Digital Humanities and the Transcending of Mortality," *Opinionator*, January 9, 2012, accessed October 21, 2013, http://opinionator.blogs.nytimes.com/2012/01/09/the-digital-humanities-and-the-transcending-of-mortality/.
31. Gaye Tuchman and Nina Fortin, *Edging Women Out: Victorian Novelists, Publishers, and Social Change* (New Haven: Yale UP, 1989), 53–54.
32. Johanna Drucker, "Humanities Approaches to Graphical Display," *Digital Humanities Quarterly* 5, no. 1 (2011). Stephen Ramsay, *Reading Machines: Toward an Algorithmic Criticism* (Illinois: University of Illinois, 2011). Dan Cohen, "A Conversation with Data: Prospecting Victorian Words and Ideas," Dan Cohen, last modified May 30, 2012, accessed May 5, 2014, http://www.dancohen.org/page/2/.
33. This shortcoming is evident in Cohen's "prospecting," with Frederick Gibbs, in the Google corpus to test the claims of Walter Houghton's classic *The Victorian Frame of Mind*: we have to take their results largely on faith, or at least on the strength of their other evidence. They therefore characterize this and other admittedly "inchoate" experiments as complements to existing research processes and eschew any claim to quantitative analysis of the kind that the Google Culturomics group made with their ngram viewer (see Jean-Baptiste Michel, Yuan Kui Shen, Aviva Presser Aiden, Adrian Veres, Matthew K. Gray, Joseph P. Pickett, Dale Hoiberg, Dan Clancy, Peter Norvig, Jon Orwant, Steven Pinker, Martin A. Nowak,

and Erez Lieberman Aiden, "Quantitative Analysis of Culture Using Millions of Digitized Books," *Science* 331, no. 6014 (2011): 176–82).

34. Matthew L. Jockers, *Macroanalysis: Digital Methods and Literary History* (Illinois: University of Illinois, 2013), 168.
35. Harris notes that the corpuses on which much text-mining work typically draws are generally skewed against writing by women. Amy Earhart demonstrates that a high proportion of early DIY recovery projects aiming to recover the works of women, people of color, the working classes, and other marginalized groups have simply disappeared as digital humanities work has become more specialized and large digital humanities grants mostly go to canonical projects. Within nineteenth-century studies, Marjorie Stone and Keith Lawson note that all the high-profile single-author projects are devoted to white, male, canonical writers (save Emily Dickinson); women get represented in groups, collectivized. We need the distanced perspective of the web of relations but need also to be able to delve into the particularities of writer or text to probe the complexities buried in larger patterns. Amy E. Earhart, "Can Information Be Unfettered? Race and the New Digital Humanities Canon," *Debates in the Digital Humanities*, ed. Matthew K. Gold (2012): 309–18. Katherine D. Harris, "Big Data, DH, Gender: Silence in the Archives?" *triproftri*, last modified March 3, 2012, accessed May 5, 2014, http://triproftri.wordpress.com/2012/03/03/big-data-dh-gender-silence-in-the-archives/. Stone and Lawson, "'One Hot Electric Breath,'" 101–26.
36. Thomas N. Friemel, "Why Context Matters," *Why Context Matters: Applications of Social Network Analysis* (Wiesbaden, DE, VS Verlag für Sozialwissenschaften, 2008), 10, doi: 10.1007/978-3-531-91184-7.
37. Eliza Meteyard, "Eliza Meteyard Letter to Leigh Hunt, June 26, 1848," transcribed by Anne Covell, Leigh Hunt Letters Collection (University of Iowa Libraries, Iowa City, Iowa), accessed April 20, 2014, http://digital.lib.uiowa.edu/cdm/search/collection/leighhunt/searchterm/Meteyard,+Eliza,+1816-1879/field/subjec/mode/exact/conn/and/cosuppress/.
38. Charlotte Yeldham, *Margaret Gillies RWS, Unitarian Painter of Mind and Emotion, 1803–1887* (Lewiston, NY: Edwin Mellen Press, 1997).
39. "The Society of Painters in Water-Colours," *The Art Journal* (London: James S. Virtue, 1861): 173, Google books edition, accessed April 28, 2014, http://books.google.ca/books?id=dD8cAQAAMAAJ&pg=PA173&lpg=PA173&dq=%E2%80%9Cquasi-classic+kind+of+art%E2%80%9D+Art+Journal&source=bl&ots=xylGem-Wa2&sig=2c3yG3o60CoVY6tt4wvqazWhGsM&hl=en&sa=X&ei=8vNmU4mXJIyYyATbkYKQBQ&ved=0CC0Q6AEwAA#v=onepage&q=%E2%80%9Cquasi-classic%20kind%20of%20art%E2%80%9D%20Art%20Journal&f=false
40. Christopher Kent, "The Whittington Club: A Bohemian Experiment in Middle Class Social Reform," *Victorian Studies* 18, no. 1 (1974): 31–55.

41. David M. Blei, "Topic Modeling and Digital Humanities," *Journal of Digital Humanities* 2, no. 1 (2012): 8–11, accessed April 28, 2014, http://journalofdigitalhumanities.org/2-1/topic-modeling-and-digital-humanities-by-david-m-blei/.
42. The meaningful links within the images that follow are visualized by OVis, a prototype network visualization tool built to explore the *Orlando* data. We are working to replace it with a similar but more flexible Web-based tool that will make it possible to share such graphs, along with links to the granular materials from which they are generated, and to combine that data with other data sets.
43. Kathryn Jane Gleadle, *The Early Feminists: Radical Unitarians and the Emergence of the Women's Rights Movement, c. 1831–1851* (New York: St. Martin's Press, 1995), 1.
44. Alexis Easley, "*Tait's Edinburgh Magazine* in the 1830s: Dialogues on Gender, Class, and Reform," *Victorian Periodicals Review* 38, no. 3 (2005): 263–79.
45. Tim Berners-Lee, James Hendler, and Ora Lassila, "The Semantic Web," *Scientific American*, May 17, 2001, accessed April 28, 2014, http://www.scientificamerican.com/article/the-semantic-web/.
46. NINES: Nineteenth Century Scholarship Online, accessed April 28, 2014, http://www.nines.org/. Bethany Nowviskie, "A Scholar's Guide to Research, Collaboration, and Publication in NINES," *Romanticism and Victorianism on the Net* 47 (2007), accessed April 28, 2014, http://www.erudit.org/revue/ravon/2007/v/n47/016707ar.html. Dino Franco Felluga, "Addressed to the NINES: The Victorian Archive and the Disappearance of the Book," *Victorian Studies* 48, no. 2 (2006): 305–19.
47. Tara McPherson, "Why Are the Digital Humanities So White? or Thinking the Histories of Race and Computation," in *Debates in the Digital Humanities*, ed. Matthew K. Gold (Minneapolis: University of Minnesota Press, 2012), 139–160.
48. Jacqueline Wernimont, "Whence Feminism? Assessing Feminist Interventions in Digital Literary Archives," *Digital Humanities Quarterly* 7, no. 1 (2013), accessed April 28, http://digitalhumanities.org:8080/dhq/vol/7/1/000156/000156.html. Liu, "Where Is the Cultural Criticism in Digital Humanities?" in *Debates in Digital Humanities*, ed. Matthew K. Gold (Minneapolis: University of Minnesota Press, 2012), 490–509. Martha Nell Smith, "The Human Touch, Software of the Highest Order: Revisiting Editing as Interpretation," *Textual Cultures* 2, no. 1 (2007): 1–15.
49. Donna Haraway, "Situated Knowledges: The Science Question in Feminism and the Privilege of Partial Perspective," *Feminist Studies* 14, no. 3 (1988): 590.
50. Haraway, "Situated Knowledges," 588.
51. McPherson, "Why Are the Digital Humanities So White?," 149.
52. The medical and other scientific fields have already begun constructing huge vocabularies of millions of terms in their knowledge domains; they

have recently been joined by information management specialists, such as librarians and philosophers. Jennifer Golbeck, Gilberto Fragoso, Frank Hartel, Jim Hendler, Jim Oberthaler, and Bijan Parsia, "The National Cancer Institute's Thesaurus and Ontology," *Web Semantics: Science, Services and Agents on the World Wide Web* 1, no. 1 (2011). The Open Biological and Biomedical Ontologies, last modified May 5, 2014, accessed May 5, 2014, http://www.obofoundry.org/. The InPhO Project, n.d., accessed May 5, 2013, https://inpho.cogs.indiana.edu/. IFLA Study Group on the Functional Requirements for Bibliographic Records, "Functional Requirements for Bibliographic Records," IFLA, last modified February 2009, accessed May 5, 2014, http://www.ifla.org/publications/functional-requirements-for-bibliographic-records. "The Getty Thesaurus of Geographic Names Online," The Getty Research Institute, n.d., accessed May 5, 2014, http://www.getty.edu/research/tools/vocabularies/tgn/index.html.

53. Amit Singhal, "Introducing the Knowledge Graph: Things, Not Strings," n.d., http://googleblog.blogspot.ca/2012/05/introducing-knowledge-graph-things-not.html.
54. Singhal, "Introducing the Knowledge Graph."
55. Susan Brown and John Simpson, "The Curious Identity of Michael Field and Its Implications for Humanities Research with the Semantic Web" (paper presented at the IEEE Conference on Big Data, Santa Clara, California, October 6-9, 2013), 77–85, accessed April 28, 2014, doi: 10.1109/Big Data.2013.6691674.
56. This despite awareness within VIAF of the complex relationships of names and identities. See Lorcan Dempsey, "Names and Identities: Looking at Flann O'Brien," Lorcan Dempsey's Weblog: On Libraries, Services and Networks, July 2, 2013, http://orweblog.oclc.org/archives/002212.html.
57. Haraway, "Situated Knowledges," 194–95.
58. Paul Baker and Amanda Potts, "'Why Do White People Have Thin Lips?' Google and the Perpetuation of Stereotypes via Auto-Complete Search Forms," *Critical Discourse Studies* 10, no. 2 (2013): 187–204. "UN Women Ad Series Reveals Widespread Sexism," UN Women, October 21, 2013, accessed April 28, 2014, http://www.unwomen.org/ca/news/stories/2013/10/women-should-ads.
59. George Eliot, *Middlemarch* (Oxford: Clarendon, 1987), 139.

CHAPTER 4

FRANCES TROLLOPE IN A VICTORIAN NETWORK OF WOMEN'S BIOGRAPHIES

Alison Booth

N.B. The images associated with this chapter are housed in the digital annex at www.virtualvictorians.org.

For the sake of inquiry, imagine a woman, Frances Milton Trollope (1779–1863), as a person in a network of the short biographies of women that circulated in books printed in Britain and North America in the nineteenth century. This persona is supposed to represent a historical individual with many documented relationships and activities. Most of the reasons for noting Frances Trollope today have to do with her role as a novelist, as the mother of writers, and as a travel writer. She gained international renown with her satiric documentary about the new republic, *Domestic Manners of the Americans* (1832), and subsequently published novels and travel narratives at an alarming rate that salvaged the family finances.[1] Yet the woman known as "Mrs. Trollope" has not become the epitome of a Victorian woman writer.[2] Her daughter-in-law Frances Eleanor Trollope's two-volume biography, published in 1895, begins: "Forty years ago, any list of Englishwomen of Letters would have been held to be strangely incomplete without the name of Frances Trollope. Fashions change; reputations fade; books are forgotten."[3] Quite so, but there are also successive reconstructions of cultural periods and the careers of the once renowned.

In 1910 Clara H. Whitmore published a precursor of feminist criticism, *Woman's Work in English Fiction from the Restoration to the Mid-Victorian Period*, a series of multisubject biographical chapters that begins with "Margaret Cavendish—Aphra Behn—Mary Manley" and ends with a joint chapter on the Brontës and a solo chapter on Gaskell. In the chapter before the Brontës, Whitmore assesses Trollope's career at some length, sandwiching the appraisal between short notices of Julia Pardoe and Harriet Martineau. The three are authors of "enduring work" at the birth of "the modern novel, with its exact reproduction of places, customs, and speech" (a863.bio14.par.1). Trollope and Pardoe are writers who "found material for stories in unfamiliar places" (par.6); Trollope and Martineau are women "of fearless honesty" (par.22). For Victorian readers, Trollope was the author not only of *Domestic Manners* but also of novels (by their short titles and date of publication): on slavery and factory labor, *Jonathan Jefferson Whitlaw* (1836) and *Michael Armstrong* (1840); scenes of clerical hypocrisy, *The Vicar of Wrexhill* (1837); bourgeois picaresque, *The Widow Barnaby* (1839) and its sequels; travels, *Paris and the Parisians* (1835), *Vienna and the Austrians* (1838), and other books. Trollope recently has enjoyed some revival of interest, alongside others on Whitmore's 1910 list who lived into the middle of the nineteenth century, such as Maria Edgeworth.[4] Although Trollope is a somewhat dim star, I have found her extraordinarily illuminating amid the constellations formed by the English-language books published 1830–1940 (excluding reference works, monographs, and periodical articles) that collect biographies of women. The example of Trollope helps to draw out the designs and potential of the Collective Biographies of Women project (CBW). The project studies the conventions and variations in the short biographies in this bibliography of books, the implicit and explicit categorization of the biographical subjects, and the kinds of networks that are constructed over time in this discourse. We use the term *documentary social networks* for the selection and co-occurrence of representations of people, to underline the anachronistic and typological construction of these cohorts.

A search for "Trollope" in the CBW database turns up Frances (P09108) and two sons: Anthony (1815–82; P18612), because he takes over one of Frances's chapters in a collective biography (a628.bio25), and Thomas Adolphus (1810–92; P14607), because he wrote a collective biography, *A Decade of Italian Women* (1859; a810), and contributed a short biography of Vittoria Colonna to *Lives of Celebrated Women* (1875; a514). Anthony Trollope, the most famous member of the family today, was the most celebrated author named Trollope in the 1880s as well. Seeking out the maternal novelist in CBW materials has brought Anthony's older brother, who was a notable nineteenth-century man of letters, out of the shadows. Thomas Adolphus Trollope (I will at times refer to him as T. A. T.) lived much of his life in Italy, keeping house with his mother, writing Italian histories and biographies, and participating in the writing projects of his wives: Theodosia Garrow (1816–65) and Frances Eleanor Ternan, sister of Dickens's mistress Ellen (1835–1913) and author of the first biography of her mother-in-law, *Frances Trollope* (1895), noted above.[5] The Trollopes in Florence, not unlike the Brownings, contributed to a Victorian revival of the Italian humanist past and the cause of independence. T. A. T.'s preface to *A Decade* repeats a commonplace of historiography at that time: "The degree in which any social system has succeeded in ascertaining woman's proper position, and in putting her into it, will be a very accurate test of the progress it has made in civilization."[6] In spite of his mother's professed conservative views,[7] her assessments of cultural differences carry forward into T. A. T.'s representation of women in Italian history and the family's contribution to a reimagined Italian nation.

The theme of the Trollope family in Italy is beyond the scope of this brief essay. Instead, I want to engage with the persona and networks formed by printed texts. A digital study of this genre of biography contributes to Victorian studies an enhanced understanding of nonfiction representations of women across the century. Victorian Britain is often viewed as a culture preoccupied with biography and self-development, yet scholarship has focused only on a small portion of the biographies published in that period. Plutarchian "parallel lives" of women burgeoned in the nineteenth century as print and literacy expanded.[8] I have found the texts in the CBW bibliography to be rich material for studies of women of many sorts, not just a literary history of women writers of fiction and poetry. Moreover, as the project develops, our interpretations of documentary social networks of personae and narratives lead in promising directions both for digital prosopography, which has tended to focus on eras before printing, and for narrative theory, which has tended to skirt nonfiction and the wide range of Victorian publications beyond the novel.

Prosopography can be defined as collective biography, of which all-female printed collections of short biographies are a subset. During the nineteenth century, however, the term in English carried two apparently disparate meanings. It denoted the representation of one person's physical traits, due to the root meaning of *prosopon* as "face"; it was also defined as the study of reconstructed, briefly sketched lives of persons in groups. Today prosopography has been reinvented through the use of computer databases, primarily among historians of early periods or cultures in which documentation is scarce.[9] As we consider the intersection of Victorian studies and digital humanities, the concept of prosopography acts as a hinge between the Victorian passions for encyclopedic knowledge as well as individualized documentation, on one side, and the Internet and social media on the other. In order to suggest the kind of analysis that CBW affords, I will align the set of books that include biographies of Frances Trollope with other emerging samples and consider further the representation of women of Italy in the database, including T. A. T.'s *A Decade of Italian Women*. But first, I will introduce the project and touch upon the challenges facing biographical databases.

CBW, Some Definitions, and Some Indeterminacies

CBW began as an online bibliography associated with my 2004 book, *How to Make It as a Woman*: documenting 1,270 books published in English that collect three or more biographies exclusively of women, it provided a vast collaborative register of changing perspectives on roles for women in many periods and countries. From 2008, in the University of Virginia's Scholars' Lab and Institute for Advanced Technology in the Humanities, with the support of the English Department, we branched into studies of the 8,600 persons and the 13,400 short narratives included in this varied genre of prosopography, or collective biography. To study the narratives, we developed a unique approach using a stand-aside XML schema, Biographical Elements and Structure Schema (BESS). BESS is a taxonomy of types of elements that editors identify and tag as occurring in the paragraphs of a narrative, in subversion, using Oxygen software. In the following list of the BESS elements, the italicized words are examples of controlled vocabulary to name the types of these elements that might be noted in a specific textual location.

1. StageofLife: *before*, *beginning*, *middle*, *culmination*, *end*, *after*
2. EventType: e.g., *crime against persona*
   a. AgentType: e.g., *male superior*, *unnamed*
   b. LocationSetting: e.g., *village*
   c. LocationStructure: e.g., *cottage*
   d. Dates, TimeofDay, Season
3. PersonaDescription: e.g., *physically daring*, *persistent*
4. Discourse: e.g., *dialogue*, *prospective or foreshadowing*
5. Topos: e.g., *influence*, *emotional effect on working men*

While many projects produce digital editions of large archives of printed texts according to the standardized markup of the Text

Encoding Initiative (TEI), and others algorithmically "mine" even larger corpuses, CBW focuses on what I call *midrange reading*, where the social norms, textual conventions, and data relating to a person's life intersect in multiple versions. Although we do not mark up the texts in the usual way, we do have sets of TEI text files on which we base our separate XML files of BESS analysis, linked to numbered paragraphs, not sentences, in each biographical chapter. Note that BESS vocabulary does not delimit actual persons, events, or other facts. The exceptions are explicit dates and overtly identified countries or continents (Ireland and France have special narrative zing from an English and eastern US perspective). The texts' linguistic patterns remain accessible for other kinds of reading, close or distant, for different purposes. When we assess all the versions of one persona, we do give unique ID numbers to unique events associated with standard date and GIS data, as far as we can trace these from authoritative biographies, such as Pamela Neville-Sington's entry in the *Oxford Dictionary of National Biography*.[10] As I will illustrate, the versions of Trollope in CBW are extremely selective, narrating few of the 28 events that, in CBW, we have named as *kernels* (at the core of the life story) or *satellites* (notable and more or less likely to be included in short accounts of Trollope's life). We compare all versions of one persona, trace the frequency and selection of events, and align these with BESS interpretation of all the biographies in a single book or across the interrelated collection types and social categories. This should lead to a model of nonfiction narrative quite distinct from existing narratology of plot and discourse in fiction. And in particular, it yields what we would never gain by writing a new biography of Trollope or any one person: measurable deviations over time in the ways this specific biographical narrative coordinates with thousands of other examples in an argument about women's lives.

Our morphology of biographical narratives is designed to remain interactive with a searchable database representing persons, texts, and historical contexts. A *sample corpus* consists of page images and TEI files of all the books in our database that include the biography of a certain woman. In documentary social networks, women who never met or communicated may be virtually associated by their proximity to each other in printed collections or may be associated through other interpretative lenses. Two persons in the same collection are called *siblings*, in one degree of separation. Two persons who never appear in the same book but share a second book with others who do appear in that first book are called *cousins*, in two degrees of separation. We began with sample corpora or sets of books formed by the fact that they include a certain nodal woman: the saintly nurse Sister Dora (1832–78) in one sample, and the adventuress known as the Spanish dancer Lola Montez (1821–61) in the other.[11] Both these Victorian women became famous under a faintly Catholic pseudonym, but their personae occupy opposite ends of the spectrum of feminine types, and they never appear in the same book; they are only cousins. Building a set of BESS profiles of the biographies in the 20 books in the Sister Dora (SD) corpus, which we call "Noble Workers," and the biographies in the 14 books in the Lola Montez (LM) corpus, "Women of the World," we turn to new sample corpora to extend our comparative study.

The Frances Trollope corpus (working nickname "FT books") is the smallest of four new sets. Trollope is a versatile figure who is *not* a sibling of either Sister Dora or Lola Montez. She is included neither among the 141 persons who appear (one to ten times each) in Noble Workers, nor among the 151 more disparate (and less recurring) figures in Women of the World, although these three sets of books were published in overlapping ranges of years: SD, primarily 1880–1913; LM, primarily 1908–1930; FT, 1883–1929. (All three samples include some outlying collections in the later twentieth century.) We plan sample corpora of Caroline Herschel, the astronomer (1750–1848), one of a handful of scientists and sisters of famous men, who is a sibling of SD but not of LM (CH is in 25 books with 268 siblings); Queen Cleopatra (32 books with 817 siblings, including LM); and Charlotte Corday (1768–93; 22 books with 351 siblings). Like Trollope, Corday, the assassin of Marat during the French Revolution, never coincides with Dora or Lola, but she virtually meets Cleopatra, Herschel, and Trollope in various publications.

Any biographical repository encounters practical as well as theoretical problems that prove more interesting upon inspection. A database pressures information to be unambiguous. What is a person—or less grandly, how do we pin down a name with authority and link all records with all variant names? Computers want events to be punctual turning points or ranges of standard dates, yet much human action consists of phases segueing gradually into one another. The key designation "male" or "female" poses problems, unsurprisingly; however binary digital code must be, the networks and narratives in CBW dislocate gender as a fixed, lifelong attribute or as a clear hierarchy. "A woman unlike other women," "braver than any man"—such expressions of the failures of categories are common in these texts. Trollope was de facto head of her household and breadwinner even during her husband Thomas Anthony Trollope's life. How common was it for a bachelor eldest son to serve as travel companion, agent, and housemate for his mother? The categories that identify people often shift questionably from supposed givens of the accident of birth (Englishwoman) to events that create family relations (mother) to activities that earn a living or renown (writer). These, for better or worse, are the terms upon which contemporaries and successors of persons in the CBW books recognized and represented them.

The names of women, especially women of rank in remote centuries or "foreign" countries, present special challenges for the encyclopedia or repository. Not only have most married women adopted husbands' names, but in general women have also been less literate and less likely to leave documented property transactions than their male counterparts. The birth dates of women, including Trollope, are more elusive than men's, sometimes because of their desire to hide their age. Variations in name "authorities" especially plague women. Women who spoke languages other than English, who were named in other alphabets, or who lived in early periods may be referred to by place of origin or family name in ways that don't match current naming practices, generating duplicate records. Many women in CBW books were associated with noble or royal families, and nothing is more bizarre and confusing than the morphing of names of persons of rank. Italian naming includes variations on the plural, masculine, or feminine versions of the family name as well as several ways to write the article meaning "of the." Whereas dukes and popes acquire numbers to distinguish them, this is not the custom with women. Disambiguating persons is one of the central tasks of prosopography, and at the same time the challenge is to interlink alternate names and meanings.

TROLLOPE IN VICTORIAN NETWORKS OF WOMEN

Within the framework of a reasonably accurate database of persons, events, and texts, the sample corpora can serve not to document a few women's historical interactions or contemporary reputations, but to measure trends and intersections in versions of hundreds of lives within categories of persona type and collection type. The scale should be zoomed out enough to get a big picture without losing sight of the form, texture, and vital facts of short narrative biographies. What do we gain from choosing Trollope as a node in a sample corpus? First, I chose her because she is a writer. By various measures, this is a popular persona type in English-language collections of biographies, though many writers are included more frequently than Trollope's 10 times: Charlotte Brontë, 46 times; Jane Austen, 27; George Sand, about whom Trollope wrote admiringly in spite of the Frenchwoman's scandalous personal life, 13.[12] Second, she is a traveler. At least since Mary Wortley Montagu's letters from Turkey (22 biographies in CBW), travel writers have gained some lasting presence, and in CBW several Victorian travelers headline this role: Mary Kingsley (6) and Isabella Bird (6) are both siblings to SD but not to LM. Trollope's biographies are triggered by her judgments of Americans, with less acknowledgment of her original social studies of European countries. Third, she is a mother. Trollope's death in 1863 did not have the common effect of inspiring a full-length biography and a quick succession of short versions; indeed, it was not until 1883 that five short biographies appeared in collections, among them Laura Carter Holloway's *The Mothers of Great Men and Women, and Some Wives of Great Men* (a411) with 36 chapters, such as "Abraham Lincoln's Mother" and "Charles Dickens's Mother." Mrs. Trollope most likely crystallized as a subject because of Anthony Trollope's *An Autobiography* (posthumously published in 1883), as well as *What I Remember* (1888) by Thomas Adolphus Trollope. I have noted his wife's 1895 *Memoir* (*FET*), which extensively quotes from the sons' records and from Frances Trollope's books and correspondence. The short biographies of Frances Trollope published before the end of World War I—the focus of this essay—tend to address fans of the late Anthony Trollope's novels. Finally, because of her writings, travels, and expatriate life, Trollope touches upon puzzling issues of nationality. Of the ten "FT books," five were published exclusively in the United States, and all focus on her role as an Englishwoman censuring American culture. From the 1820s, the Bristol native formed lasting social connections in France (General Lafayette and his family were devoted to her), Austria, and especially Florence, Italy, where she lived most of the time from 1843 to 1863, events all but missing from the CBW biographies. No one dreams of calling her anything but English.

Collections have been organized around each of the above types or categories that made a biography of Trollope narratable and worth including in various collective biographies.[13] I do not mean to reinforce the idea of identity as a fixed label—quite the opposite. The frames of reference overlap, and the catalogues of representation within documentary social networks elicit contiguity and analogy, not transparent identification. Most recorded historical women—including many of the queens and wives, mistresses, mothers, or sisters of notable men—can be termed "learned women" or "writers," and any traveler is probably known as such because she wrote about her voyages. Individuals break molds but they also personify them, making the role newly viable. The presenters who wrote and assembled these biographies adjusted to audience expectations but also had to pay some respect to documented events that did not conform to a stereotyped prescription for women's lives.

BESS analysis of women's lives in documentary networks reveals trends in representation beyond the awareness or agency of individual authors, publishers, or audiences at the time. We uncover rhetoric and principles of selection, bearing in mind that there is an element of the arbitrary or the interchangeable in prosopographies of women. Trollope intersects in her sample corpus with some 193 individual women, including Cleopatra, Caroline Herschel, and Charlotte Corday, in ten collections of short biographies of women. (She is one of 25 persons in the database who recur ten times.)[14] Why is Trollope *not* in one of the SD or LM sets? Trollope was considered a good family woman who served social causes, within reach of the readers' experience; a biography of this writer would not seem out of place among Noble Workers, but she is absent, perhaps largely because of locations and themes of nationhood. She provoked a longstanding resentment mainly among American readers, whereas Sister Dora figured primarily as a Victorian local saint who resembled her contemporary Florence Nightingale. In contrast, Trollope would seem out of place among the splashy, exotic, and sometimes ephemeral lives in Women of the World. Although she traveled widely, held a salon, and associated with nobility and celebrities, she was more bourgeois than bohemian, a wife and mother rather than a "beauty," neither a performer nor an adventuress. The notorious Lola Montez belongs in a tabloid framework quite opposed to industrious literary production that saved the family fortunes.

We can learn a great deal from playing the types and their cohorts against the known biographical data about the actual women. None of the nodes of our sample corpora ever actually met each other. For a short period in the later eighteenth century, Herschel, Corday, and Trollope were contemporaries. (Trollope was in her teens when Corday was guillotined in Paris in 1793.) Again, after Sister Dora's birth in 1832 (the year that Trollope's *Domestic Manners* hit it big as a salvo in the debate about reform), there was a brief span of contemporaneity for Trollope, fellow British women Lola and Dora, and the Hanover-born Herschel, who came to fame with her brother in Bath and Windsor. In the 1830s Trollope was hoping to take her daughter to visit family friends in Hanover (which from 1714 to 1837 was under the rule of English monarchs of the House of Hanover) (*FET* 157, 168). In 1848, the same year that Herschel died (aged 98) in Hanover, Montez, still in her twenties, was fomenting revolution in Munich, some 390 miles to the south. Sister Dora, whose brother was the famous Oxford don Mark Pattison, lived in Yorkshire and Staffordshire, and her brief career as nurse and hospital administrator flourished in the decade after Montez ended her globe-trotting career in New York in 1861. Meanwhile, Frances and Thomas Adolphus lived in Florence, in a house they called "Villino Trollope." The mother—the other nodal figures are considered childless (Herschel, Sister Dora, and Charlotte Corday are seen as maiden spinsters, Montez and Cleopatra as serial lovers or bigamists)—died in 1863, probably unaware of the then-obscure Sister Dora. Although Trollope traveled and provoked scandal, not unlike Montez, she is more like Harriet Beecher Stowe or Elizabeth Gaskell (well-traveled married mothers who wrote about social issues) than the Women of the World. Such comparisons suggest the ramifying connections that we trace in the CBW project: among events and places, among the documentary social networks constructed by the printed books and the database, and among BESS analyses of narratives.

The texts associated with Frances and Thomas Adolphus Trollope provoke anxiety about national character, and they offer instances of the intersectional and sliding significations of identity, including name and gender-normative role. Questions of nationhood haunted the importation of Italian and French catalogues of women over the centuries. Nineteenth-century collections, at times claiming to describe universal womanhood, quite often stress national character, as in "Engl" (England or English, 64 [numbers = books in CBW]); "British" (12); "Britain" (8); "America[n]" (88). Certain national types stand out, sometimes sorted by epoch and geography. Frenchwomen, prominent not only in society but also in arts, letters, and politics, pervade the database.[15] Joan of Arc is the most frequent subject of all, with 68 short biographies, but other Frenchwomen who confronted monarchs or populace include Marie Antoinette (45) and Empress Josephine, the wronged wife of Napoleon (29). France fostered the *salonnières* of the seventeenth century as well as heroines of the Revolution or the Empire, including Madame Roland (35) and Mrs. Trollope's friend, the celebrated beauty and *salonnière*, Madame Récamier (15). Women of Italy produced a more muffled and delayed effect on female prosopographies in English (10 titles refer to Italy), though they tended to flourish at least a century before the famous women of the French court.[16]

When is Italy?

As I approached the Frances Trollope sample corpus and thought about her son's collective biography, *A Decade of Italian Women*, I decided to explore my longstanding interest in nationality (influenced by GIS and spatial trends in digital humanities). How did Italian women fare, alongside the French and other nationalities? Compared to the BESS interpretation of narratives, I expected to find national identification in CBW to be relatively straightforward. I had presupposed that the event of birth helped to fix this attribute, which should have some consensus among existing sources. Yet data about places and events, including publication, create an interactive and shifting picture.

Nationality has developed into its modern association with birth certificates, passports, and the nation-state, but it has long been changeable, whether through migration or such political transformations as the fall of the Soviet Union. A person's nationality, beyond the geographical location favored by computation, turns out to be as flexible as his or her vocational role. Biographies in any form, print or online, customarily give some indication of geography, above all the places of birth and death, along with residence and occupation. Many life narratives note the event that BESS vocabulary tags as *move to new country*, or other actions that can be named with our terms regarding travel and home. Short biographies of Trollope are full of the topoi related to *national character*. The topos type *patriotism* underlies passages in the American-authored biographies of Trollope, whereas it never surfaces in the BESS analyses of biographies of Sister Dora, who ignored contemporary politics and scarcely left northern England. We need the term *patriotism* in the story of Lola Montez only in the episodes, in Poland and Munich, when her enchantment of a ruler provoked revolutionary unrest—though she was a cosmopolitan whose girlhood was in India and who took Europe, New York, California, and Australia by storm.

For the presenters of collective biographies, as for us in our work on the CBW database, Italian women required more research and perhaps more extenuation to bring forward. The ladies of Ferrara or Venice or Florence could rival their male contemporaries in learning. But women of Italy were more notorious even than those of the French court, and their historical remoteness meant that their status as sexual objects in the few extant depictions of them could go unchallenged (as if the Duke of Ferrara in Robert Browning's "My Last Duchess" had the last word). Italian women indelibly figure in the history of art, as the subjects of portraits. Canonical poetry, too, by Boccaccio, Dante, Petrarch, and Tasso among others, immortalizes women who once walked in Italian cities. Some Italian women, such as Dante's Beatrice or Petrarch's Laura, shone as a chivalrous ideal, the muse and beloved of a great man.

To determine our list of Italians, an assistant quite sensibly asked, should we include Romans? No, the compilers of biographies distinguish the classical era from the revival of classical learning in that same region. Further, many individuals whom we would fit into our picture of the Italian Renaissance were born or died beyond the borders of modern Italy, depending on various alliances among royal or noble houses across Catholic Europe. In the seventeenth century, for example, Cardinal Mazarin imported to the French court a set of his "nieces," heiresses who married into the upper echelons of French society; their story figures prominently in France, though one collection (a870) calls it *Five Fair Sisters: An Italian Episode at the Court of Louis XIV* (by Hugh Noel Williams).[17] Renée or Renata, Duchess of Ferrara, to give another instance, is also known as Renée of France, daughter of Louis II. What should we do with a woman of Savoy? After its feudal dynasty, it came under French rule, then joined the Kingdom of Piedmont-Sardinia, then became part of France in 1860. And those Florentines got around: Marie de' Medici (1575–1642) was the second wife of Henry IV of France and mother of Henrietta Maria of England. The attribute of nationality thus is debatable for any biography, perhaps especially so for celebrated women, and all the more so for women associated with the kaleidoscope of principalities that sought to become Italy during the Trollopes' residence in Florence.

Thomas Adolphus Trollope was not alone as an English writer interested in women of Italy, but was notable for regarding them as precedents for a resurgence of female influence in letters, the arts, and political leadership. The 410 pages of volume 1 of *A Decade of Italian Women* present the following, in multiple chapters each (IDs and number of collections are indicated here)[18]:

- St. Catherine of Siena, 1347–1380, who influenced the return of the Pope from Avignon to Rome (P09198; 12)
- Caterina Sforza, 1463–1509, a warrior queen in Antonia Fraser's 1988 prosopography (p036A), a ruler of strategic alliances (and assassinations) among papal and ducal parties (P10232; 3)
- Vittoria Colonna, 1490–1547, scholar and poet (P09955; 9)

Volume 2, 450 pages long, proceeds to more recent women—less written about, and treated in fewer words in T. A. T.'s collection:

- Tullia d'Aragona, 1510–1556, called "the Intellectual Courtesan" (P11500; 4)
- Olympia Morata, 1526–1555, celebrated writer and Protestant convert (P13668; 11)
- Isabella Andreini, 1562–1604, actress and playwright (P08484; 3)
- Bianca Cappello, 1548–1587, Venetian mistress and later wife of Francesco de' Medici, Duke of Tuscany; they were believed to have been poisoned together (P08679; 7)
- Olympia Pamfili, 1591–1657, heiress of Maidalchini, sister-in-law of Pope Innocent X, powerful leader in Rome (P14612; only in *Decade*)
- Elisabetta Sirani, 1638–1665, painter (P13859; 2)
- La Corilla, or Maria Maddalena Morelli, 1740–1800, a self-made court poet and improvisatrice crowned at the Capitol in Rome in 1776, a model for de Staël's Corinne (P14614; only in *Decade*)

T. A. T. found out some of his own subjects—the more recent ones. In the CBW list, two of his subjects are unique. Significantly, he follows a surge in interest in the prominent figures in volume 1, who became popular biographical subjects less than 20 years before: Vittoria Colonna and St. Catherine of Siena, for example, appear in no collective biographies in CBW that were published before 1830. Though Trollope's cited sources are Italian, he would have had precedents in decades of magazine articles, from *The Foreign Quarterly Review* in 1830 ("Bianca Cappello," by Fortunato Prandi) to *The Dublin University Magazine* in 1858 ("Vittoria Colonna," anonymous). It is fitting that T. A. T. wrote of these historical women in a household of women writers and among advocates for the Risorgimento.[19] And it is fitting that he wrote in Florence: it seems likely that he thought of La Corilla because "in the Via della Forca, at Florence, the eye of an observant traveler may remark a marble slab let into the front of an otherwise undistinguished house, and bearing the inscription 'Here lived La Corilla in the eighteenth century.'"[20]

Other professional writers specialized in Italian topics, rivaling Trollope. Edgcumbe Staley introduces a747, *Famous Women of Florence* (London: Constable; New York: Scribner's, 1909): "One of the most brilliant tokens of the *Vita Nuova* of Civilisation—which we call, perhaps incorrectly, 'The Renaissance'—was undoubtedly the New Woman" (xi). He complains that Bianca Cappello is too often "stigmatized" "as a courtesan. Perhaps Thomas Trollope, in his 'Decade of Italian Women,' has transgressed the bounds of truth and decency more glaringly than any of them. I make the strongest protest possible against his falsehoods!" (xx). Trollope was on a similar mission to Staley's, but in 1859 he couldn't have met the "New Woman." T. A. T. is puzzled that a period of debilitating and "truly ignoble wars" was also a time of flourishing intellectual achievement for women (*DIW* 1:291–92). He wonders at Lucrezia Borgia's transformation into the good Duchess of Ferrara with her fourth marriage, after committing "abominations wholly unreproducible in any modern page" (*DIW* 2:38). He writes that Lucrezia's "restoration and *rehabilitation* . . . would be impossible in the nineteenth . . . century," which proves "the enormous amount of advance in the moral sense of mankind" (italics in original; *DIW* 2:41). Inevitably, the Protestant North is the moral vanguard, but it has lost the vitality of the Renaissance (a period T. A. T. helped to name). Undoubtedly, "the great intellectual movement in Italy in the sixteenth century" sponsored "celebrated . . . poetesses," such as Tullia d'Aragona (*DIW* 2:1). The present age could study the "constellations of such stars" as Vittoria Colonna in "that fascinating dawn," or spring, "of European civilisation." What new social conditions might enable Victorian women to move toward "the increased solidarity, co-operation, and mutual influence of both the sexes?" (*DIW* 1:vi–vii). But then T. A. T. backs off from "speculations on 'the woman's question'" as he presents "his little cabinet of types of womanhood" in relation to "social environment" (*DIW* 1:viii). For English and American sojourners in Italian-speaking cities during the 1840s–1870s (many of whom visited Villino Trollope), a history of Italian women and the current events of the Risorgimento formed a diptych of aesthetic and political appeal. Witnessing the upheavals of 1848 and 1859 and beyond, T. A. T. observes that the British no longer merely regard Italy as an art museum but have "begun to awaken . . . to the fact, that there is on that sunny side of the mountains a live and struggling nation with high aspirations."[21]

The representation of Italian women seems to have been a strategy, perhaps veiled even from its proponents, to shape a nation in the image of expatriate Florence: an advanced state of civility in which women and the arts flourish. If Theodosia Trollope's articles influenced the British outlook on the Risorgimento, Thomas's biographical histories seem to design a related effect. *A Decade of Italian Women* is hardly radical; many of its pages consist of gentlemanly discussion of conflicts within the church and among princes. The very word "Italian," however, was a feat of historical imagination and a hopeful slogan in 1859. A "decade" (a set of ten) echoes the ten days of collected tales of *The Decameron* and implies that a series of historical biographies can also represent a transformative epoch. But Italy was an ongoing event during the Trollopes' residence. The vital inspiration of the place (beyond the state) and its women (beyond the usual roles) could be recovered as well as envisioned.

FORESHORTENED LIVES, SUBLIME NETWORKS

Frances Trollope resided in Italy for almost as long as did the fifteenth- and sixteenth-century notables on her son's list, with their cruelly short lives. But of course she doesn't belong in that cohort. She is a cosmopolitan traveler, not an immigrant or a native—a member of expatriate literary circles and not a beauty or poet at court. Catherine de' Medici (1519–89), who is also a French queen, does appear in one FT book, Abbot's *Notable Women of History* (a001), a copious list of 73 such women. But only three other Italian-born women show up in FT collections, each once only. These are the two singers included in Abbot (a001)—Adelina Patti, born in 1843 when Trollope moved to Italy, and Adelaide Ristori, whom she saw perform (*FET* 97)—and the Princess of Belgiojoso (in Adams's *Celebrated Women Travelers of the Nineteenth Century* [a007]). We could analyze each of the FT books and their persons by their geotemporal and other data, as I have begun to do with *A Decade of Italian Women*. Here I indicate some other ways to evaluate these materials. This list gives the flavor of persona types in eight FT books, bearing in mind that the same person may occupy more than one role and that some of these persons appear more than once as a sibling of Trollope:

- 1 South Asian
- 2 adventurers
- 2 Jews
- 2 painters
- 2 saints
- 3 nurses
- 3 scientists
- 4 Italians
- 5 mistresses
- 7 classical figures
- 12 performers

- 12 political/military figures
- 14 contemporary reformers
- 15 nobles/aristocrats
- 16 travelers
- 17 wives
- 20 queens
- 26 Frenchwomen
- 28 Americans
- 31 mothers
- 55 writers

Comparing the documentary networks in this way enriches our understanding of the reception of the persona's life: how is her narrative valued and deployed?

Certainly a collection of writers, travelers, or mothers will tie the narrative to its central theme. But I have been surprised by the extent to which a writer's biography, even in this short form, becomes *literary analysis*, a discourse type that I had to add to our BESS vocabulary when I dug into the FT material. The norm in the biographies that we have analyzed so far is to pace along the chronological track (with foreshadowing and retrospection to comment on the persona's achievements) and to mention all the known phases of the life. Usually, male and female biographers adopt similar tones or stances toward the persona. The seven versions of Trollope (one is reprinted in three books) diverge from both these patterns. Three twentieth-century collections aim at accurate recovery of intrepid women (the authors are Una Pope-Hennessy a650 [1929]; Helen Heineman p044A [1983], who published a Twayne study of Trollope in 1984; and Barbara Hodgson p045A [2002]). These female authors acknowledge a pioneering writer: one who in 1827 briefly joined Fanny Wright's utopian community on the Mississippi River; one who, accompanied by three children and Auguste Hervieu, a French artist, moved to booming Cincinnati and tried to launch America's first combination shopping mall and theme park;[22] and one who went on to write celebrated books. Like many feminist studies recovering women in the past, they foreground private life, noting her marriage, her six children (three of whom died of tuberculosis), and her thriving literary salon in Florence. Clara Whitmore, whose 1910 study of women writers I noted above (a863), also views Trollope through a feminist lens but curtains private life according to the convention of literary studies of an oeuvre. Another female biographer, Laura Carter Holloway, contributing to the spate of FT biographies in the 1880s, chooses Trollope as one of "the mothers of great men" (in her book of the same name [a411]) and like Whitmore devotes paragraphs to a range of Trollope's books rather than her life story; surprisingly, it is the mothers' claim to greatness, not the maternal relationship, that Holloway's chapters seek to establish, and that claim for Trollope is based in published works. The texts by male presenters—William Henry Davenport Adams in 1883 (a007.bio12), Willis J. Abbot in 1913 (a001.bio70), and James Parton in 1883–88 (a624.bio26, a626.bio03, a628.bio25)—would never be mistaken for one of the feminist versions of a hundred years later. The American-based Abbot and Parton reduce Trollope to an incident in a cultural quarrel between England and America that measures more than half a century of American progress. Adams, an Englishman ("our business . . . is with Mrs. Trollope the traveler" [par. 28]), takes the trouble to read closely or sample Trollope's "clever" writings and ascribes to her a "masculine mind" (par. 8, 26), but his tone is cavalier.

The FT books reinforce my finding, as we built the bibliography, that titles are unreliable. Holloway's chapter 23, "Frances Trollope the Mother of Anthony Trollope," begins with the question of the recently dead son's "greatness." "As both the brothers inherited their literary abilities from their mother, Frances Trollope, who was born in 1780 and died at Florence in 1863, she is in our judgment entitled to a place among the mothers of great men" (a411.bio23 pars. 1–2). Thus compressing the mother's birth and death into a single sentence and justifying her inclusion in the volume, Holloway has little further use for Anthony (he is only mentioned as "a refined and modified edition of his mother" [par. 14]) and never represents parenting or family relations. Holloway's topic is "Mrs. Trollope as an author" of "quantity rather than quality" (par. 4). "The day has, we trust, passed away when Mrs. Trollope is to be judged by her first book . . . It would be as unfair to sentence her to literary death because of her attacks upon our manners . . . as it would have been to refuse to read Charles Dickens's later works" because of the mockery in *American Notes* (par. 3). (The discourse type *direct address, use of we* is exceptionally frequent in Trollope's biographies.) Holloway nevertheless focuses paragraphs 14–19, including more than four pages of paragraph 15, on Trollope as a satirist and her outrageous, culminating first book. The chapter concludes, "Mrs. Trollope's looking-glass has two sides" (par. 19).

A wide variety of biographies in CBW digress into biographies of the men in the women's lives and thereby into the conflicts over power, territory, and belief that make traditional histories. Thus Parton's biography (entitled "Mrs. Trollope") begins with the word "Cincinnati," names Mrs. Trollope in paragraph 2, never narrates her birth or life before coming to America, becomes a biography of Anthony Trollope for paragraphs 30–43, and wraps up with this perfunctory concluding paragraph (44): "Mrs. Trollope died in Florence in 1863, aged eighty-three years. In private life she was a very friendly and good soul, much admired and sought in the society of Florence, where she passed the last twenty years of her long life." The English-born Parton, whose first wife was the writer Fanny Fern, hardly seems interested in this English female novelist; his theme is Anglo-American relations and the changing conditions of American society. He devotes five paragraphs to the Trollope family's encounter with President-Elect Andrew Jackson in Cincinnati "on his triumphal journey to Washington" (22–27). Indeed, Parton had written a full-length biography of Jackson (1859–60).

For American biographers, Trollope is essentially one event, that book, and a verb, *to Trollopize* (that is, to criticize harshly or abuse). Trollope's short biographies, partial in both senses, seem indifferent to the task of creating an accurate précis of the subject's life, quite unlike the biographies in the cohorts of Sister Dora, Lola Montez, and Caroline Herschel that we have analyzed so far. T. A. T. mustered as much as was known about his Italian women, but like his mother they often survived merely as a controversial episode and a handful of writings by or about them. Chapters on Trollope downplay most of the beginning and end of her life.[23] The biography of the American astronomer Maria Mitchell that precedes the chapter on Trollope in Parton's *The World's Famous Women* (a626.bio02), for example, narrates the entire span of Professor Mitchell's life and appears to have been written by someone who had been her student at Vassar.[24] The following chapter, by contrast, can only imagine a tale of the happy Trollope family in Cincinnati: "Fresh from England, and retaining all their English love of nature and out-of-door exercise, the whole family, parents, two sons and two daughters, often climbed" the bluff on the edge of town (a626.bio03 par. 1). Parton builds his chapter on the notion that Anthony shared in all his mother's observations of the New World. But in fact, one son, Henry, journeyed to Cincinnati, and Mr. Trollope and Thomas Adolphus only joined the family later. Other biographers also have little reliable evidence; Adams asserts in his first sentence that Frances Milton "was born at Heathfield Parsonage in Hampshire, in 1787" (a007.bio12 par.1), whereas according to the *Oxford DNB*, it was Heckfield in 1779. Abbot's sketch, "Frances

Trollope (1782–1863), Whose Book on America Enraged a Nation," begins,

> being fifty years old, never having written a line for publication, but being in the American idiom "stone broke," it occurred to Mrs. Thomas Anthony Trollope, an English woman of good family, to recoup her fortunes by writing a book . . . Her qualifications in the way of observation were that she had spent three days in New Orleans and rather more than two years in Cincinnati, with brief visits to New York and Washington, about 1830. By way of fitting her to be judicial and fair she had lost her last dollar in a "bazaar."

That's an age for foolhardy ventures, but no other version confirms 1782; most archival records, including the Library of Congress, believe Trollope was born in 1780.[25] Misinformation is one thing, but bias is another, and Abbot's is clear enough. Amid censure of Trollope's hostility, the scanty PersonaDescription in these versions seems preoccupied with defining the class and gender of her textual persona rather than providing a detailed characterization of the woman herself: "Mrs. Trollope has been described by her friends as a refined woman of charming personality. But as soon as she began to write, she donned her armour and proclaimed her hostility" (Whitmore a863.bio14 par. 11). Adams is more severe: "The great fault of Mrs. Trollope is, that she is always a critic and never a judge . . . It must be admitted that she is often vulgar" if a "shrewd observer" (a007.bio12 par. 1). Satire, Victorian biographers contend, is masculine and unladylike as well as unaesthetic. Holloway equivocates on the topic of Trollope's offense; she "overdid" the satire (par. 9), but

> it is nonsense to call her unfeminine; a social satirist has no sex, and if she criticised the life around her merely from a feminine standpoint, her observations would be far less trustworthy than when she does so as an intelligent human being . . . She shows no mercy; but the set of people she took off deserved no mercy. Although she was a consistent Tory all her life, her political satires cannot be called untruthful or unjust. (par. 14)

This sort of characterization of an authorial persona is familiar in *literary analysis*, with its persistent use of present tense, otherwise rare in biographies analyzed to date.

According to these narratives, the persona herself does very little other than write and publish a lot after a belated debut, a rather unstimulating tale (we note *pause* and *gap* in narration). The StageOfLife outlines of these texts show *culmination* and *after* (that is, posthumous events) in the same paragraphs as the 1832 *publication* (E00082), followed by her anticlimactic career, which can only be associated with the iterative stage, *middle*, which in many biographies is a tally of recurrent occupational practice rather than distinct scenes that shape the plot of the life. Yet the biographers can't resist emulating and excerpting her witty descriptions of society and culture, which seem to be as familiar to readers of these biographies as a novel by Dickens (who is frequently mentioned). Three male biographers of men—Parton, Adams (a London-born writer included in the *ODNB*), and Abbot (a prolific American journalist and biographer)—cast Mrs. Trollope as uninformed and prejudiced, yet they take part in a political quarrel that she started as they sample or rewrite her texts. Trollope is a precursor of sociologists, ethnographers, and bloggers; she is among the few middle-class women in Britain or the United States who sparked a long-lasting debate on national issues, though she had little to do with abolition or women's suffrage. BESS work on Trollope texts called for new "topos" vocabulary, *conservatism* and *liberalism*, and featured unusually high rates of vocabulary related to *national character* (topos) as well as discourse types *description of a kind of people or society* and *periods or times compared*. We expect comparable patterns of narration and theme to appear in some biographies of writers, travelers, politicians, or missionaries, where one encounter or intervention (or oeuvre), as well as national controversy or ethnography, will tend to overshadow personal narrative.

Such observations are in progress as we assemble a new sample corpus and will truly become telling when we analyze all the siblings, cousins, and outliers in six or more sample corpora, across decades and types of collections and personae. A new interface, improved data and functionality of the database, complete repositories of text images and files, as well as BESS interpretations and visualization tools—all are under way toward releases in coming years and freely accessible meanwhile. We hope to collaborate with users who wish to approach individuals and groups, as well as those who want to adapt the BESS approach to other studies of narrative. In addition to linking our database with archival authorities on persons, we foresee using different methods to yield networked patterns in this genre (including topic modeling). Genealogists, prosopographers, and massive aggregators of reference and documentation on persons and texts, such as Wikipedia and WorldCat, might well tap into the potential for parsing the trends in printed biographies of men and women, in short or full-length form. We think human curation can help us shed more light in the near term on a distinctive and neglected genre, and can yield more nuanced interpretations of narrative structure and ideology.

A prosopographical database alters one's attention to individuals, blurring historical or textual specificity. At the distant scale of national and international archives, ontologies wobble. The binaries of encoding and the vagaries of names, nationalities, and other data of individual lives present challenges for a study of changing representation of women and of narrative genre. Reading midrange, between close and distant reading, does not mean that we are not inspired by the Niagara of textuality that we can now quantify and paint in dazzling colors. Parton quotes Trollope's description of "'four days of excitement and fatigue at Niagara,' where, as she says, 'we drenched ourselves in spray, we cut our feet on the rocks, we blistered our faces in the sun, we looked up the cataract and down the cataract, we perched ourselves on every pinnacle . . . we dipped our fingers in the flood. . . .' In all these delights the future novelist had his part. Let us hope, too, that he shared with his parent the pleasure she took in the Hudson River" (a624.bio26 pars. 39–401). Indeed, let us perceive British and American biographical history in terms of networks of shared stories, including counterfactuals about persons who were never on the scene together in actual space and time. Anthony Trollope did not frolic with his mother in the sublime landscapes of America.

NOTES

1. Frances Trollope, *Domestic Manners of the Americans* (London: Gilbert & Rivington, 1832). "Such literary fecundity is terrifying," writes Willis J. Abbot in *Notable Women in History* (London: Greening, 1913), the first book in the bibliography of Collective Biographies of Women, http://womensbios.lib.virginia.edu. All references to such texts will be cited parenthetically by their identification number, chapter number (e.g., bio70), and paragraph numbers. In the cited passage, Abbot continues: he claims a London obituary (unidentified) compared Trollope's productivity to "the reckless production of children." Mrs. Trollope's 115 volumes call for a "union for the regulation of hours and of output" (a001.bio70 pars. 9–10).
2. Like Trollope, Adelaide Procter appears in ten books, including three with Trollope. No biography of M. E. Braddon enters this database. Margaret Oliphant appears 4 times; Elizabeth Gaskell, 7; Christina Rossetti, 7; Elizabeth Barrett Browning, 37.
3. Frances Eleanor Trollope, *Frances Trollope: Her Life and Literary Work from George III to Victoria*, 2 vols. (London: R. Bently and Son, 1895). Cited henceforth parenthetically as *FET*. See Juliette Atkinson,

"Fin-de-Siècle Female Biographers and the Reconsideration of Popular Women Writers," in *Writing Women of the Fin de Siècle: Authors of Change*, ed. Adrienne E. Gavin, Carolyn W. de la L. Oulton, and Linda H. Peterson (New York: Palgrave Macmillan, 2012), 111–23.

4. See for example Tamara Wagner, "Frances Trollope [Special Issue]." *Women's Writing* 18, no. 2 (May 2011): 153–292. Biographies of Trollope include Johanna Johnston, *The Life, Manners, and Travels of Fanny Trollope: A Biography* (New York: Hawthorn Books, 1978); Helen Heineman, *Frances Trollope* (Boston: Twayne, 1984).
5. Frances Eleanor Trollope writes that both Thomas and Anthony wrote autobiographically about their mother, "yet she appears only by a side-light" (vii). The biographer warmly praises her mother-in-law's character: in addition to "the most flawless sincerity, and a warmly affectionate heart" (*FET* 1:55), she had a dedicated domestic nature (61, 100) and was also an accomplished writer whose salon entertained all distinguished visitors to Florence (2:54).
6. Thomas Adolphus Trollope, *A Decade of Italian Women*, 2 vols. (London: Chapman and Hall, 1859), http://archive.org/details/decadeofitalian-w00trol. 1: v. I will cite this work parenthetically: (DIW *1: v*). Although it is a "CBW book," a810, the text is not part of our sample corpora and not available through our site.
7. "Despite all the new liberalism, and the rational sympathy which she felt for the cause of Italian freedom, Frances Trollope always retained a grateful and pleasant memory of her stay in Austria" (*FET* 2:255). See also *FET* 2:105.
8. For discussion of the history of the genre, see Alison Booth, *How to Make It as a Woman: Collective Biographical History from Victoria to the Present* (Chicago: University of Chicago Press, 2004). Of course there are antecedents for the Victorian collections in hagiography and humanist catalogues, as in the pre-1830 chronological list at Collective Biographies of Women http://womensbios.lib.virginia.edu/browse?section=1. Translations of Scudéry's, Brantôme's, and others' compilations of *femmes fortes*, or illustrious ladies, influenced English productions, and in 1752 George Ballard's *Memoirs of Several Ladies of Great Britain: Who Have Been Celebrated for Their Writings or Skill in the Learned Languages, Arts and Sciences* seemed to announce a "Renaissance" in Britain.
9. Alison Booth, "Prosopography," in *The Wiley-Blackwell Encyclopedia of Victorian Literature*, forthcoming. See "Who Were the Nuns?" http://wwtn.history.qmul.ac.uk/index.html and "The Continental Origins of English Landowners, 1066–1166," http://www.coelweb.co.uk/coeldatabase.html, among many other digital prosopographies.
10. "Trollope, Thomas Adolphus (1810–1892)" and "Trollope, Frances (1779–1863)," Pamela Neville-Sington in *Oxford Dictionary of National Biography*, ed. H. C. G. Matthew and Brian Harrison (Oxford: OUP, 2004); online ed., ed. Lawrence Goldman, May 2008, http://www.oxforddnb.com.proxy.its.virginia.edu/view/article/27751 (accessed May 28, 2014).

11. Lola's associates include some who share pages with Trollope: Queen Elizabeth I, Catherine the Great, Zenobia Queen of Palmyra, Jenny Lind, George Sand. Lola Montez or LM books tend to adopt performers, Spaniards, beauties, courtesans, and political entrepreneurs: Madame Récamier; Sarah Churchill, Duchess of Marlborough; Nell Gwyn; Peg Woffington. Italians Bianca Cappello and Tullia d'Aragona ("the intellectual courtesan") surface in LM lists and in *A Decade of Italian Women*.
12. Trollope singles out Sand from the disreputable mob of French writers, "for genius is of no nation" (*Paris and the Parisians*, Letter LXIII, 340). Marie-Jacques Hoog, "Trollope's Choice: Frances Trollope Reads George Sand," in *Woman as Mediatrix: Essays on Nineteenth-Century Women Writers*, ed. Avriel H. Goldberger and Germaine Brée (Westport, CT: Greenwood, 1987), 59–72. If Sand was not to be tied down to nationality, she was also exceptionally difficult to tether to gender, with her cross-dressing, pseudonym, and sexual freedom.
13. Of the 1,270 collections in CBW, at least 85 books announce that they represent writers; 17 are devoted to historical mothers, among 32 titles referring to "mothers"; and 10 collect biographies of travelers.
14. Currently, 208 "persons" are listed in one degree of separation from Trollope (P09108), but some of these in fact are not individual women. For a usable total of 193 individuals, I did not count the two post-1940 collections, which favor group chapters on themes; one of these books is organized by geography, so Trollope's life appears in several chapters. Toru Dutt (1856–77) is apparently the only woman of color in FT books, though the topics of racial difference and abolition surface in versions of her life.
15. We find 43 titles that announce a group of Frenchwomen or mention "France"; more than 123 Frenchwomen add historical texture and flavor to these books. We have identified 136 Italian persons, of whom 11 are male.
16. In the Pop Chart http://womensbios.lib.virginia.edu/popchart (rating names appearing four or more times in samples across three periods), 17 of the 113 names are French, but only 2 are Italian: Catherine of Siena and Vittoria Colonna. Trollope intersects frequently with names in the Pop Chart: 65 out of 113 women are also siblings in FT books.
17. A search for "five" in the title field under Books/Collections, http://cbw.iath.virginia.edu/public/books.php, produces a list of interesting variants on "five" and nationhood, such as a912, *Five English Consorts of Foreign Princes*; others by ID numbers include p024; a281.
18. These can be searched by entering part of each name in the "full name" field on the Persons search page, http://cbw.iath.virginia.edu/public/women.php.
19. Theodosia Trollope translated Italian literature and published articles in *The Athenaeum* that were later collected as *Social Aspects of the Italian Revolution* (1861). Thomas Trollope, for his part, weighed in on the

liberal side of the Italy question in the short-lived *Tuscan Athenaeum* and other journalism, as well as in historical articles and such books as *Tuscany in 1849 and 1859.* T. A. T. and Frances Eleanor Trollope collaborated on *Homes and Haunts of the Italian Poets* (1881).

20. Thus begins the chapter entitled "The Apprenticeship to the Laurel" (*DIW* 2:393). The Italian-language Wikipedia site shows an image of the house, with the plaque in capital letters: "Qui Abitò Corilla Nel Secolo Decimo Ottavo."
21. Thomas Adolphus Trollope, *Filippo Strozzi: A History of the Last Days of the Old Italian Liberty* (London: Chapman & Hall, 1860), http://catalog.hathitrust.org/Record/008640645. Quoted in Lawrence Poston, "Thomas Adolphus Trollope: A Victorian Anglo-Florentine," *Bulletin of the John Rylands University Library of Manchester* 49 (1966): 133–64. 143.
22. Mr. Trollope and T. A. T. visited the family in Cincinnati briefly in 1828, Anthony remaining at school, but the US sojourn and business venture were mostly in Mrs. Trollope's hands.
23. To compare versions of a persona, we name key events (standard date, GIS), adding ID attributes in the BESS files. Events that would be kernels or that one would expect to be nearly ubiquitous are often missing in biographies of Trollope: Birth (E00079) is missing in three versions; Death (E00088) is missing in two; and we have seen the cursory way these facts are dealt with in two versions. Only one version deals with Trollope's move to Italy, and only three note her major novels. Her marriage or her husband's death, her important travel writing other than *Domestic Manners*, and the births and deaths of her children are all elided in four or five out of the five versions studied.
24. Her classes and soirees are described in the present tense. The discursive chapter headings in a626 frequently begin with birth and end with death; in Abbot's collection, too, subjects besides Trollope have their life-spans traced in detail (a001).
25. LC n 50012547 says, "1780-1863." Ditto VIAF, ID: 34551336 http://viaf.org. I follow the *ODNB*, which cites the Bristol Record Office.

CHAPTER 5

REPRESENTING LEIGH HUNT'S *AUTOBIOGRAPHY*

Michael E. Sinatra

In his biography of Leigh Hunt, Anthony Holden asserts, "Alongside Wordsworth, who largely eschewed literary London, Hunt's was the longest nineteenth-century literary life, with the widest circle of acquaintance and as large a claim as any to the shaping of literary opinion" (2). In my earlier monograph, *Leigh Hunt and the London Literary Scene*, I illustrated the kinds of change that Hunt's reputation went through over a 30-year timespan. That study attempted to elaborate the problematic of his position within the London literary and political scene between the years 1805 and 1828, the contributions he made to British literature and journalism, and his public standing at the end of the romantic period. Since Hunt's life is obviously too complex to be rendered fully in any single study, the idea was not to attempt an exhaustive history, but rather to present a starting point for further inquiry into Hunt's career as a writer and public figure under the reign of Queen Victoria.

The trajectory of Hunt's life and work traced in the years 1805 to 1828 appears to fall off rather steeply toward the end, leaving Hunt as a failed author, with neither a dependable, receptive audience nor any clear future as an author in sight as the third decade of his public life comes to an end. To conclude this account of Hunt's early career abruptly at this point, however, would be to create a misleading impression of his career as a whole. Hunt survived the setbacks of *The Liberal* and *Lord Byron and Some of his Contemporaries*, and he remained more or less true to his principle of independence. The second half of his life was very fulfilling for Hunt, both personally and professionally. Hunt's later years deserve more attention from twenty-first-century scholars, and they offer many topics of research worth pursuing further. During that time, Hunt made new friends and lost most of his enemies; his advancing age also had an impact on the stamina he could invest in new literary projects, such as running a periodical or going to the theater every night. His reputation as a well-known and respected personality grew during these years and brought him many visitors from around the country as well as from America, where his works had been regularly published and reviewed since *Juvenilia*, in the same way that those of Coleridge and Wordsworth had been between 1830 and 1850. Yet all throughout the second half of his life, Hunt retained a keen political sense and a sharp critical judgment, which motivated him to promote new poetical voices, such as Tennyson and D. G. Rossetti, as well as past authors, in his critical anthologies and editions *Imagination and Fancy* (1844), *Wit and Humour* (1846), *Stories from the Italian Poets* (1846), and *Beaumont and Fletcher* (1855).

My essay focuses on a reading of the complex politics of authorial revisions and the reception history of Hunt's 1850 *Autobiography*, with some reference to *The Examiner* since this is arguably Hunt's most famous publication as editor and the work that gave him a place of prominence within the romantic period.[1] The importance of independent judgment, first expressed in Hunt's earlier work in journalism in *The News*, clearly influenced Hunt's journalistic career in *The Examiner*, as the prospectus for this periodical (included at the back of *Critical Essays* and reprinted in the first issue of *The Examiner*) demonstrates. This lengthy advertisement also serves as further evidence of the sociopolitical implications of Hunt's theatrical criticism, the way in which *Critical Essays* constitutes the first instance of Hunt's longstanding insistence on independence from external pressures (whether they be editorial, personal, or political), and his early engagement in social and cultural issues. Hunt links critical and political independence in his choice of motto for *The Examiner*: "Party is the madness of many for the gain of a few." He thus effectively advertises the impartiality of his new periodical by using his current reputation as an impartial theatrical critic:

> The Gentleman who till lately conducted the THEATRICAL DEPARTMENT in the NEWS, will criticise the Theatre in the EXAMINER; and as the Public have allowed the possibility of IMPARTIALITY in that Department, we do not see why the same possibility may not be obtained in POLITICS. (Hunt, 2003 1, 31)

Hunt distinguishes *The Examiner* further by describing the contemporary tendency of the press: "The newspaper proves to be like the generality of it's species, very mean in it's subserviency to the follies of the day, very miserably merry in it's puns and it's stories, extremely furious in politics, and quite as feeble in criticism" (Hunt, 2003 1, 31). Hunt asserts in the conclusion of the prospectus that, just as he had cleared the way for a new, unbiased drama criticism when he started writing for *The News*, so too would he change political journalism and provide a new, neutral voice within the contemporary press with the founding of *The Examiner*.

The Examiner rapidly rose to success, and the sales were very strong in the first decade of the newspaper's existence, with a circulation of approximately 2,200 issues by November 1808, rising to a peak of between 7,000 and 8,000 in the 1810s. These circulation figures are quite impressive when one bears in mind the limited numbers of copies sold by all the publications of that period; for instance, the *Edinburgh Review*'s circulation was 12,000 and *The Times*'s 8,000 (Deguchi, 1996 vii). This success can be ascribed in part to the shared commitment to reform of both Hunt and his brother John, and in part to Hunt's personality as editor. Indeed, Jeffrey N. Cox and Greg Kucich rightly explain that part of the success of the newspaper, in terms of both longevity and influence, comes "from the power of Hunt's writing, which is by turns chattily erudite and aesthetic, cleverly satirical, and filled with political rage" (Hunt, 2003 1, xxxvii).

The Examiner played a major role in the London political scene, as well as in the literary periodical world. The new weekly also had an important impact on Hunt's life and career. As Kenneth Neill Cameron notes, "*The Examiner* became not so much a weekly paper as an institution and Leigh Hunt was transformed from an obscure poet and essayist into an influential editor, a man whose opinions were read and admired by thousands of readers week by week for some thirteen years" (1961–70 1, 263). Hunt's periodical came to have a major influence on an entire generation of writers in the early decades of the nineteenth century, from a political as well as a literary perspective. Studies devoted to Shelley, Keats, and Hazlitt frequently include a discussion of Hunt's newspaper, since it played such an important role in their writing careers. *The Examiner* also provides modern readers with the proper contextual information for Keats's and Shelley's poems, as Nicholas Roe and Cameron, among others, have persuasively argued.

On June 8, 1850, the three-volume edition of *The Autobiography of Leigh Hunt; with Reminiscences of Friends and Contemporaries* appeared under the imprint of Smith, Elder, and Co. Although based in large part on works previously published, Hunt's *Autobiography* is probably the most important work of his later life. While a large section of the material included in Hunt's *Autobiography* comes from *Lord Byron and Some of his Contemporaries*, the tone is greatly altered. Whereas *Lord Byron* was very much a statement of personal justification in the face of the various attacks Hunt had suffered in publications on Byron, and a reaction against the fulsome praises that the dead poet now garnered from around the country, the *Autobiography* offers a calmer depiction of Hunt's life, imbued with an obvious sense of pleasure in the recollection of past events and friendships. Hunt's revision pleased reviewers, for several either allude or explicitly refer to Hunt's gentler treatment of Byron. For example, *The Palladium* notes Hunt's more benevolent treatment and reproduces passages of "the apologetic remarks" (138); *Harper's New Monthly Magazine* similarly marks the difference between 1828 and 1850 by noting that the later work shows "the asperities of his nature gently worn away, and his mind brought under the influence of a kindly and genial humor" (572). The *Methodist Quarterly* highlights Hunt's revisions the most of these three reviews, asserting that, for Hunt, "the chief delight which he enjoyed in writing his own life seemed to result from the opportunity afforded of setting forth motives once misconstrued, and expressing manly regret for early indiscretions" (253).

As Ken A. Bugajski notes in his article "Editing and Noting: Vision and Revisions of Leigh Hunt's Literary Lives," "Even Hunt's best and most successful periodical endeavor, the *Examiner*, does not escape revisionary criticism in 1850" (n.p.). In Hunt's 1828 volume, *Lord Byron and Some of his Contemporaries*, he had described the founding of *The Examiner* thus:

> At the beginning of the year 1808, my brother John and myself set up the weekly news paper of the Examiner in joint partnership. The spirit of theatrical criticism continued the same as in the News, for several years; by which time reflection, and the society of better critics, had made me wiser. In politics I soon got interested . . . I was very much in earnest in all I wrote . . . I think precisely as I did on all subjects when I last wrote in it. (411)

When Hunt reworked this passage into the 1850 *Autobiography*, he significantly changed its tone in recasting the verbatim section he used in a more negative light:

> At the beginning of the year 1808, my brother John and myself set up the weekly news paper of the Examiner in joint partnership. It was named after the Examiner of Swift and his brother Tories. . . . I thought

> only of their fine writing, which, in my youthful confidence, I proposed to emulate . . . I wrote, though anonymously, in the first person, as if, in addition to my theatrical pretensions, I had suddenly become an oracle in politics. . . . I blush to think what a simpleton I was. . . . The spirit of the criticism on the theatres continued as it had been in the News. (2, 1–3)

Thus, Hunt now "views the name of the periodical as an arrogant usurpation; he discounts the certainty of his political beliefs; and his widely respected theatrical opinions have become pretension" (Bugajksi n.p.).

Interestingly, just as Hunt's work after 1828 tends not to be considered by modern critics, Hunt himself is curiously silent about the later part of his life. To some extent the rather abrupt truncations of the account of his literary career may be due to the practical necessities of producing copy under pressure of time. The process of revision was certainly cut short by Smith, Hunt's publisher, who insisted, in a new contract dated February 7, 1850, that Hunt should produce the manuscript within three months.[2] Stephen Fogle comments on Hunt's financial motivation for publishing his autobiography in 1850 when he asserts that "the circumstances of the composition of the book, that is, the need to make good on his contract . . . go far to explain this emphasis [on Hunt's early life]. Much of the material lay ready to his hand, suitable for reprinting once the rights were cleared" (vii–viii). Indeed, the pressure to meet his contractual obligation with the firm in time may also have been a motivation for Hunt to borrow heavily from his previous publications, principally *Lord Byron and Some of his Contemporaries*, the essays on Italy he published in *The Liberal*, and some articles from *The Examiner*.[3] However, Hunt may also have been attempting to recuperate his stance of independence by revisiting and revising this earlier controversial material into more temperate terms. Since it was in the early period of his life when he was most fiercely independent, it is that period that he spends the most time revisiting.

Thomas Carlyle's enthusiastic reaction to Hunt's *Autobiography*, in a letter dated June 17, 1850, is one of the most positive comments that the volume gathered after its publication:

> I call this an excellently good Book; by far the best of the autobiographic kind I remember to have read in the English Language; and indeed, except it be Boswell's of Johnson, I do not know where we have such a Picture drawn of a human Life as in these three volumes. A pious, ingenious, altogether *human* and worthy Book.[4]

If the reviews of Hunt's book were generally mixed, they were nevertheless very numerous: the *Autobiography* received no less than 15 reviews in Britain and Ireland, and 5 in America. The style of Hunt's *Autobiography* is heavily criticized, particularly for what the reviewers feel to be dullness, in the *North British Review*, *The Palladium*, and the *Dublin University Magazine*.[5] The anonymous reviewer for *The Spectator* acclaims Hunt's *Autobiography* as an enriching source of information on literature and society,[6] and the anonymous reviewer for *The Times* observes that Hunt's life is an interesting subject for a book, even though the financial problems that figure so prominently argue for an origin in pecuniary motives rather than an interest in literary history.[7] Hunt's central place in the London literary scene is recognized favorably in the reviews for *Tait's Edinburgh Magazine* and for *The Literary World*,[8] and Hunt's appreciation of his fellow authors is commended in *Chamber's Edinburgh Magazine* and in *The Literary Gazette and Journal of Belles Lettres*.[9] Alongside Smith's offer to publish Hunt's recollections, these reviews attest to Hunt's widely respected position within the London literary scene and the considerable interest in his memoirs.

Hunt's choice of material for discussion in his *Autobiography* might suggest that he never considered himself to be part of what is now called the Victorian period. It might also simply indicate that Hunt was primarily interested in reflecting upon what he thought of as the best years of his life, including of course his friendships with Keats and Shelley. In any case, one can legitimately question whether Hunt should be considered as a Victorian autobiographer as well as a key romantic figure whom modern periodization tends to overlook. His life and success under the reign of Queen Victoria complement his pivotal role during the romantic period, and his *Autobiography* contains much undiscussed material that is relevant to both literary periods. Nevertheless, Hunt's *Autobiography* does not give, to adopt Anthony Trollope's words, "a record of [Hunt's] inner life,"[10] and one needs to turn to his 1853 book, *The Religion of the Heart*, to find a detailed expression of Hunt's personal beliefs, as I have explored elsewhere.[11] Commenting on the revisions Hunt made in the *Autobiography*, Timothy Webb—in the words of Bugajski—"argues that [they] result from a conscious change in philosophical perspective through which Hunt begins to look at fellow humans more charitably. Webb writes, for example, that Hunt's 'gradual process of revisionary evolution . . . strongly suggests that for Hunt the process of revision was not only a matter of stylistics or even of truth to history and to self but to an activity whose deepest resonances were moral and religious'"

(Bugajski n.p., quoting Webb 299). The religious sentiment was not, however, always welcomed by contemporary readers, as demonstrated by the anonymous reviewer for *The Palladium* who criticized what he perceived as anti-Christian sentiments in Hunt's *Autobiography* (137) and by a similar anonymous complaint that appeared a year later in *The Rambler, a Catholic Journal and Review of Home and Foreign Literature, Politics, Music and the Fine Arts* (47).

For inclusion in the planned Leigh Hunt Archive website, I will prepare an edition of Hunt's *Autobiography* (both the 1850 and the 1860 editions); this online resource will also include an annotated collection of all Hunt's critical writings, as well as a selection of primary and secondary works by and about other writers involved in his literary circles. These works will include letters by Charles Cowden Clarke and Vincent Novello, Benjamin Robert Haydon's diary, William Hazlitt's essays in the *Round Table* and other writings he published in Hunt's periodicals, a selection of John Hamilton Reynolds's and Charles Lamb's contributions to the *London Magazine*, and a biography of Madame Vestris, along with several other memoirs and critical writings on drama. All these works will provide a unique intertextual reading environment for the digital version of Hunt's critical writings. Furthermore, the Leigh Hunt Archive website will be constructed along the lines of Jerome McGann's *Rossetti* archive, "so that its contents and its webwork of relations (both internal and external) can be indefinitely expanded and developed" (McGann n.p.).[12] The idea is to have a "central text hypermedia"[13]—an electronic edition of Hunt's entire critical corpus, with appended notes and hypertext links, along with links to historical-critical editions. This website will allow for an exploration of a new facet of Hunt's critical productions, and it will contribute to an ongoing effort to consider the true importance of so-called minor literary figures—as well as to further study of the 1830s, a decade that (as Richard Cronin notes) does not constitute a literary period but instead gets lost between two others.[14] Since students and scholars alike need access to primary texts for their work, the site will be crucial in making possible a proper reevaluation of Hunt's writings in the first half of the nineteenth century. It will also encode Hunt's own literary networks by featuring biographical notices of other writers within his literary circles, along with reviews, notices, and a detailed chronology.

Digital editions, due to the affordances and constraints established by the compound platforms of the modern computer and the World Wide Web, are learning commons, parliamentary hubs, urban squares within which active processes of scholarly debate and exchange can be rapidly broadcast, recorded, collected, preserved, and shared. If digital editions are to take full advantage of their environments (rather than simply emulating print traditions), they need to visibly include both process and product, and to offer opportunities for editorial diligence, contribution, perspective, control, and debate to their knowledge-community of users. The Leigh Hunt Archive intends to incorporate the work of Stéfan Sinclair on data-mining tools (specifically the implementation of his *Voyant* tools in the various electronic editions prepared during the course of the project) and Jon Saklofske on visualizing data (specifically through his *NewRadial* prototype).[15] Data mining offers many opportunities to bring together different sets of data which, when prepared to the highest standard of text encoding, can yield new and innovative results that encourage further reconsideration of preconceived notions regarding the transfer of ideas from one author to another, or one literary genre to another. Furthermore, the results of the research undertaken in the Leigh Hunt Archive will be presented in a collaborative, visual context that reimagines the digital scholarly edition as a transparent workspace layer in which established primary objects from existing databases can be gathered, organized, correlated, annotated, and augmented by multiple users in a dynamic environment that also features centralized margins for secondary scholarship and debate. *NewRadial* is a site for the generation of social editions and for a more public and open process of edition formation, pluralization, and persistent growth. It is also a site of scholarly process, discussion, and development. It has the ability to represent database material in a sandbox environment, thus encouraging iterative experimentation, hosting methodological and interpretative debate, and supporting new juxtapositions and connections. Most important, adapting *NewRadial* for specific use with this project will place the Leigh Hunt Archive database in conversation with efforts relating to the idea of the semantic web. *NewRadial*'s use of an RDF (Resource Description Framework) data model to organize and export any secondary scholarship that grows out of user interactions with its primary database makes it extremely useful for prosopography- and placeography-related data manipulation, and for compatibility potential with other RDF-oriented applications such as NINES.[16]

Ultimately, the Leigh Hunt Archive will be useful for anyone working on the romantic and Victorian periods because it will provide access to important contextual information that allows for a better understanding of the key literary and historical events between 1800 and 1850. The biographical notices that it will include—along with a planned series of recorded interviews of scholars discussing Hunt and the other authors under consideration, as well as relevant information on topics such as the methods of publications available at the time and the evolution of Hunt's literary circles—will make this website an important resource for researchers, students, and the public at large. It will feature contextual information useful for anyone interested in (say) freedom of the press, the rise of the historical novel and historical plays, antiwar poetry, or autobiographical writing. The website will make these texts available for the first time in electronic format; what's more, the site itself, through the use of Geographic Information System (GIS), will generate customizable visual maps of London's literary circles and their contributors. These will allow users to explore the encoded material in original and innovative ways that extend beyond Hunt himself and yet reassert his centrality to the romantic and Victorian periods. Thanks to the data-mining and visualization tools to be implemented in the project, the Leigh Hunt Archive will feature cutting-edge methods for searching and analyzing the large body of data that will have been scanned and prepared to the Text-Encoding-Initiative standards (thus ensuring full compatibility with other electronic resources as well long-term preservation and accessibility). Mass-digitization projects such as Google Books do not offer researchers the same level of granular searches or visualization tools, and thus there is still the need for such a database to be constructed from the ground up.

Thinking of editorial representation today leads one to consider the complex relationship between digital humanities and literary studies, bearing in mind that, with its emphasis on tools, digital humanities can seem to be detached from traditional literary methods even though it arguably became prominent thanks to its origin in literary studies. Howard Besser asserts in the 2004 *Companion to Digital Humanities* that

> though the promise of digital technology in almost any field has been to let one do the same things one did before but better and faster, the more fundamental result has often been the capability of doing entirely new things. (558)

An interdisciplinary field that before the Web appeared to specialize in electronic concordances, digital humanities is now training students in disciplines ranging from philosophy to history to communicate through the Web and use its powerful resources. Further, digital humanists are working with libraries to develop the electronic archives that are the durable research content that scholars use to understand themselves and their history. The fundamental restructuring of the research record represents a vast modernizing opportunity that is a necessary step forward given the ever-increasing dominance and enabling features of the digital media.

A large number of digital humanities projects have grown out of, or found a happy home in, digital humanities centers around the world. As Neil Fraistat puts it in the collection of essays *Debates in the Digital Humanities*, edited by Matthew K. Gold:

> digital humanities centers are key sites for bridging the daunting gap between new technology and humanities scholars, serving as the cross-walks between cyberinfrastructure and users, where scholars learn how to introduce into their research computational methods, encoding practices, and tools and where users of digital resources can be transformed into producers. Centers not only model the kind of collaborative and interdisciplinary work that will increasingly come to define humanities scholarship; they also enable graduate students and faculty to learn from each other while working on projects of common intellectual interest.[17]

I believe that digital humanities centers and some large-scale digital infrastructures are indeed the best way forward, but what does this mean for literary studies more specifically?

The first place is probably the most obvious since it comes from the origins of digital humanities, or humanities computing, as it was more commonly called in the 1950s when Father Roberto Busa started working with IBM to produce an index to the works of Thomas Aquinas. In other words, digital humanities was arguably at first, and some would suggest still is, a set of tools that can facilitate some aspects of scholarly work by using large-scale computational processing. In that sense, digital humanities is only another step in the technological developments that have gone hand in hand with literary scholarship over the last few hundred years as we moved from orality to the technologies of literacy. Thus, far from fearing it, we should accept its potential as a new resource, one that accompanies new forms of editorialization, such as the shift from manuscripts to print editions set by hand and then to those mass-produced by the steam press.

Digital humanities also does offer new reading and annotating tools; some are already implemented (think of the shared annotation in Kindle books), and some try to break away from the skeuomorphic transfer from print format to electronic format by introducing dynamic tables of contexts (rather than contents)[18] or new methods of annotation, as Ray Siemens and his SSHRCfunded MCRI project,

Implementing New Knowledge Environments, have been prototyping for the past few years.

Lack of physical support for literary content, in the form of primary or secondary sources, is now common. This should not in itself be a source of concern for literary scholars either (except for those also interested in the history of books), since they tend to focus their analysis on the content of the work, not whether it's published in octavo format. The argument of a democratization of knowledge is one that should support the happy marriage of literature and digital humanities. Indeed, which author doesn't dream of reaching a wider audience among the readers who may look for or simply stumble across the millions of books made available by Google in the last few years?

Yet there is the argument that new ways of reading, as described by the notion of "distant reading" whereby scholars can analyze millions of books for patterns, also correspond to the end of literary studies as we know it, in that attention to details (the traditional method of so-called close reading) gets lost in the overwhelming amount of data now available. It is worth bearing in mind, however, as Eric Hayot suggests, that

> the first thing to say is that distant reading is not really distant, and close reading is not just close. No reading practice ever maintains itself as one "distance" from a text; rather what we call a reading practice is among other things a pattern of systems of habitual distances and relations among those distances. So "close reading" is not always close; rather it pairs a certain kind of analysis of relatively small pieces of text with very powerful analytic tools—the tools of New Criticism, but also of psychoanalysis, deconstruction, new historicism, and so on—that leverage those small pieces of text into structures that are more "distant" from the text than is, say, the sentence or the phoneme.[19]

Thus, digital humanities as a method of reading is once again simply another way of dealing with data—"literature as the site for the *storage l* of information," as Hayot puts it—that retains the same intrinsic quality and interest as other literary methods, namely, the pursuit of new ways to explore and understand meanings present in texts that are at the center of our scholarly investigations by retrieving information from the texts studied. In fact, Jerome McGann's latest book, *A New Republic of Letters*, quite neatly adds to this discussion:

> We see this in and as the emergence of the digital humanities, which both its promoters and its critics regard as a set of replacement protocols for traditional humanities scholarship. But the work of the

> humanist scholar has not changed with the advent of digital devices. It is still to preserve, to monitor, to investigate, and to augment our cultural life and inheritance. (4)

If the textual infrastructure in a digital world is to match the quality of textual data in the print world, scholars need to take a much more active curatorial role. Forms of "scholarly crowdsourcing"—comparable in some ways to the practices of *dispersed annotation* in genomic research—offer the promise of creating models of data curation that will maintain fundamental primary data and incrementally improve them over time. This is a big task, with progress measured in decades rather than years. It poses important technical challenges for developing new forms of man-machine interaction. It raises institutional questions of where to locate repositories and how to manage workflows and issues of quality control. It also underscores how digital humanities has grown from being understood as a tool for a range of disciplinary-based projects to a transdiscipline in itself, including at this point in its history competing definitions of its very meaning. Finally, it raises questions about the "prestige economy" of the academy and the way in which scholarly labor is ultimately allocated and rewarded.

NOTES

1. This essay borrows some data from my chapter on *The News* and takes its premise from the epilogue of my book *Leigh Hunt and the London Literary Scene* (Routledge, 2005).
2. The arrangement between Smith and Hunt is at the Brewer–Leigh Hunt Collection and is referred to in Landré vol. 1, p. 253.
3. J. E. Morpurgo's edition of Hunt's *Autobiography* contains a very useful appendix that identifies all the passages in the *Autobiography* that are reproduced verbatim from earlier sources; among these, *Lord Byron and Some of his Contemporaries* is the main instance. See *Autobiography*, pp. 496–98.
4. *The Collected Letters of Thomas and Jane Welsh Carlyle, Vol. 25—1850*, eds. Ian Campbell, Aileen Christianson, and Hilary J. Smith (Durham: Duke UP, 1997), 97.
5. Anon., "*The Autobiography of Leigh Hunt*," *North British Review* 14 (November 1850): 165; Anon., "[Review of *Autobiography*]," *Palladium* 1 (August 1850): 137; Anon., "Leigh Hunt," *Dublin University Magazine* 36 (September 1850): 272.
6. Anon., "Leigh Hunt's *Autobiography*," *Spectator* 23 (June 22, 1850), 593.
7. The anonymous reviewer also advances that Hunt was responsible for his own problems, rather than society ("The Autobiography of Leigh Hunt," *The Times* 20585 [September 4, 1850]: 7).

8. Anon., "Autobiography of Leigh Hunt," *Tait's Edinburgh Magazine* 17 (September 1850): 571; Anon., "Review of *The Autobiography of Leigh Hunt*," *Literary World* 7 (September 14, 1850): 210.
9. Anon., "[Review of Leigh Hunt's *Autobiography*]," *Chamber's Edinburgh Magazine* 14 (July 13, 1850): 23; Anon., "Leigh Hunt and His Contemporaries," *Literary Gazette and Journal of Belles Lettres* 1745 (June 29, 1850): 437. The anonymous reviewer for *The Literary Gazette* also notes that age has improved Hunt's style and predicts that the work should be popular outside the literary world (437).
10. Anthony Trollope, *An Autobiography*, ed. David Skilton (London: Penguin, 1996), 232.
11. See my essay "'A Natural Piety': Leigh Hunt's *The Religion of the Heart*," *Allen Review* 19 (1998): 18–21.
12. Jerome McGann, "The Rationale of Hypertext" (1995) http://www2.iath.virginia.edu/public/jjm2f/rationale.html.
13. Lynette Hunster, "Hypermedia Narration: Providing Social Contexts for Methodology." *Conference Abstracts*. Association for Literary and Linguistic Computing / Association for Computers and the Humanities Conference, April 1992; quoted in Claire Lamont, "Annotating a Text: Literary Theory and Electronic Hypertext," in *Electronic Text: Investigations in Method and Theory*, ed. Kathryn Sutherland (Oxford: Clarendon Press, 1997), 60.
14. See Richard Cronin, *Romantic Victorians: English Literature, 1824–1840* (New York: Palgrave, 2002).
15. Sinclair's *Voyant* tools can be found at http://voyant-tools.org; Saklofske's *New Radial* is at http://socrates.acadiau.ca/courses/engl/saklofske/newradial.html.
16. I am grateful for Jon Saklofske for his many suggestions regarding this project.
17. Neil Fraistat, "The Function of Digital Humanities Centers at the Present Time," in *Debates in the Digital Humanities*, edited by Matthew K. Gold (Minneapolis: University of Minnesota Press, 2012), 281.
18. For more on this topic, please consult the work of Susan Brown and Stéfan Sinclair/Geoffrey Rockwell in their CFIfunded projects Canadian Writing Research Collaboratory and the Voyant Tools suite.
19. Eric Hayot, "What Is Data in Literary Studies?" (January 14, 2014), http://erichayot.org/ephemera/mla-what-is-data-in-literary-studies/.

Chapter 6

Visualizing the Cultural Field of Victorian Poetry

Natalie M. Houston

N.B. The images associated with this chapter are housed in the digital annex at www.virtualvictorians.org.

The Victorians have always already been virtual. They are constructed out of our narratives of historical change, out of our interpretation and interpellation of their material artifacts, and out of the cultural residue—ideas, tropes, and images—that still circulates today, making them both oddly familiar and familiarly strange. New Historicism and post-structuralism reminded us that we can never reach the Victorians themselves (or their texts as they knew them); instead we can only access representations, filtered and processed into a flickering simulacrum of historical actuality. This virtual Victorian reality is generated by our canons, syllabi, and library collections—engines that enable us individually, and collectively as a profession, to create what seems to be a fully realized literary-historical understanding. Its terrain is limited in scope and resolution by the data (texts, images, objects) ingested by the hermeneutic process. If you have read only five Victorian novels in an undergraduate course, your personal virtual Victorian reality will be less detailed, less fully realized, than that of a well-read specialist. This is obvious. But as the collective virtual Victorian reality created and used by scholars, archivists, and curators becomes an ever-better simulacrum, it is easy to lose sight of the mechanisms that generate it. Our understanding of the Victorians depends upon resources that are made available by acts of selection.

Many humanist disciplines, including literary studies, have traditionally been suspicious of data and quantification for overly reducing complexity and producing deceptive notions of objectivity.[1] Thus the kinds of data already available within our objects of study go largely unnoticed. Rather than setting data analysis in opposition to traditional humanist reading, I want to bring them together in what I call *digital reading*, in which humanist research and interpretation draw on computational analysis in order to surpass the human limits of vision, memory, and attention.[2] In this essay, I explore two familiar forms of humanist data: the anthology table of contents and the library cataloging record. Treating these informational structures as data sources intentionally defamiliarizes them and removes them from their usual contexts of pedagogic and readerly utility. (Throughout this discussion, I focus on printed books, although of course tables of contents exist for periodicals and other codex forms, and library catalog records exist for numerous kinds of textual and media artifacts.) The mode of digital reading I demonstrate in this essay involves transforming quantitative information with visualization techniques in order to better understand two versions, past and present, of the cultural field of Victorian poetry. These cultural fields, constituted differently in the items cataloged in research libraries and in the poems gathered in anthologies, create our virtual Victorian reality.

THE CULTURAL FIELD

The field of culture, according to Pierre Bourdieu, "is a network of objective relations (of domination or subordination, of complementarity or antagonism, etc.) between positions," such as those occupied by material or symbolic entities.[3] In the literary field, these might be books or their features (such as genre or theme), agents (authors, publishers), or institutions (schools, prize committees, and so on). These different positions in the field make possible different "position-takings," such as an action, an attitude, or the creation of a specific literary work. The positions and position-takings of entities in the field produce the prestige, beauty, and other symbolic values associated with cultural objects. At any given moment, Bourdieu suggests, there are multiple fields within the space of culture that compete and change at the metafield level, just as agents do within subfields. Thus there is a field of poetry, within a field of literature, within a field of cultural production more broadly conceived. The field offers a model for understanding the material and symbolic production and reception not only of particular works, but also of the very idea of the work and of the categories through which it is understood by both writers and readers.[4]

As Peter McDonald suggests, such sociological analysis shifts "the burden of analysis from celebrated individuals or works to the objective conditions that make particular ways of writing and reading possible. . . . The primary task, then, is to reconstruct the field."[5] Such analysis might, for example, explore how Christina Rossetti's 1862 volume, *Goblin Market and Other Poems*, was situated in the field of poetry: with what other books published by Macmillan might it be associated? How many books of poetry were issued by other publishers in the same year? How was Macmillan situated in the cultural field? Although there are excellent studies of Rossetti's poetry and publishing practices, as well as histories of Macmillan, such questions require us to look more broadly at the landscape of poetry's production and circulation at midcentury. In "The Field of Cultural Production," Bourdieu admits that reconstructing these historical spaces of possibility can be difficult "because they were part of the self-evident givens of the situation . . . and are therefore unlikely to be mentioned in contemporary accounts, chronicles or memoirs."[6] The same holds true for the contemporary moment's ongoing reception of the Victorians. One way to overcome this difficulty is to investigate some of the data we overlook because of its "self-evident" functional usage.

VICTORIAN DATA

A tremendous amount of data from the nineteenth century is currently available, particularly since the Victorians themselves were interested in data collection, systems of classification, and statistical analysis. Much of the material of interest to literary scholars exists in textual forms, such as books, periodicals, letters, and memoranda, belonging to writers, editors, publishers, and general readers. In recent years, digital editions and large-scale digitization projects have made much of the published material of the Victorian period more accessible through digital surrogates. This digitization of the Victorian cultural record also makes these materials available for computational analysis.

But having a lot of texts to read (with human eyes or computational algorithms) does not in itself mean you have data. Data does not exist in the world but is instead created through our decision to pay qualitative or quantitative attention to particular artifacts, phenomena, or sources of information. Johanna Drucker suggests that humanist inquiry should understand all data as *capta*, as something "taken" rather than simply "recorded and observed": such a shift "acknowledges the situated, partial, and constitutive character of knowledge production, the recognition that knowledge is constructed, taken, not simply given as a natural representation of pre-existing fact."[7] Scholars and librarians frequently create such capta with books: we create accession and catalog records that describe physical and conceptual features of the items; we count the number of books held in a collection or the number of entries in a bibliography; and we photograph, scan, or microfilm books for preservation, access, and research. Through these activities, detailed information is created about the physical item being cataloged, including bibliographic metadata (such as author, title, publisher, date) as well as descriptive information about its physical structures (such as the size and number of pages, or the presence of illustrated plates). Moreover, the very decision to collect, preserve, or study a particular book depends upon its position within already-established systems of classification, such as the Library of Congress subject headings or the definitions that govern rare-book collecting. Long before large-scale digitization, we were creating humanist data, particularly in the form of bibliographic, biographical, and institutional metadata, information that describes and inscribes a particular text within systems of knowledge.

The forms and structures of the data thus created are already embedded in hermeneutic operations, as Lisa Gitelman and Virginia Jackson suggest:

> At a certain level the collection and management of data may be said to presuppose interpretation. . . . Data need to be imagined *as* data to exist and function as such, and the imagination of data entails an interpretive base.[8]

Traditional literary methods of interpretation might attend to a book and understand its textual contents as form, genre, and idea, while the methods associated with corpus linguistics might extract type and token counts and parts of speech, because the disciplinary contours of interpretation shape the horizon of expectations we bring to the objects of study.[9]

EXPLORATORY DATA ANALYSIS

In his foundational book, John Tukey defines *exploratory data analysis* as "looking at data to see what it seems to say" and as "detective work" that seeks to discover clues indicating where further analysis should be conducted.[10] Although graphic displays of quantitative information are frequently used to support the presentation of research findings, they can also be used as part of an initial stage of research. Exploratory data analysis adopts an attitude of curiosity, a willingness to discover what lies within the data, whether it conforms to previously held assumptions or not. In a similar vein, Frederick Gibbs and Dan Cohen argue that "one of the most productive and thorough ways to do research . . . is to have a conversation with the data in the same ways that we have traditionally conversed with literature: by asking it questions."[11]

One of the best ways to ask questions and look for clues is through data visualization, which harnesses the pattern-recognition powers of the human brain in order to enable comparison, differentiation, segmentation, and other ways of looking at complex data sets. Tukey emphasizes the intuitive understanding that visualization offers, suggesting that "it may not be possible to find words to express what a display or plot may show."[12] Today, many different kinds of plots can be created with the data-visualization tools included both in basic spreadsheet software and in more sophisticated statistical packages. The complex, multicolored, visual displays produced by such tools are best suited to viewing on-screen, not only for enhanced color reproduction, but also to enable the viewer to explore the image with zoom and pan controls.[13]

FROM THE LIST TO THE FIELD

The list is an integral information structure in academic culture: from the syllabus to the bibliography, lists of authors and texts stand in for concepts, themes, events, and ideas considered essential in delimiting fields of knowledge. Debates about curriculum design and the status of the literary canon tend to focus on the lists of authors and works included in anthologies as condensed synecdoches for these volumes (just as the name of an author frequently serves as a synecdoche for an entire oeuvre).[14] Representing literary history as a list of author names allows scholars to compare lists from anthologies or syllabi to what John Guillory calls the "imaginary totality" of the canon, created not by "a really existing list, but by retroactively constructing its individual texts as a *tradition*, to which works may be added or subtracted without altering the impression of totality."[15] This tendency to represent literary history through a list of names is deeply engrained within academic conventions and institutions.

The list is also the primary data structure of the scholarly bibliography and the library catalog: we encounter a sequence of information about each item, whether ordered by accession number, author name, title, or year of publication. Such data structures are excellent for recording information about individual items and for specifying their location in order to facilitate retrieval. The lists returned from search and browse operations in library catalogs are not, however, designed to facilitate understanding of the relations among items. At their best, today's digital interfaces to library catalogs and bibliographic databases provide multiple levels for sorting the list, such as first by author name and then by year of publication, but each item's information remains distinct.

Both the library catalog record and the anthology table of contents locate specific information in conceptual and material space: the hierarchy of subject headings frames the topic of a cataloged item within a constrained universe of knowledge, while the item's cataloging number indicates its physical location on a shelf. A table of contents instantiates a system of organization and subdivision that defines levels for perceiving meaningful items, so that poems may be indicated individually in the anthology's table of contents, as well as inscribed within a larger work, an author's oeuvre, or a time period. Each line in the anthology table of contents locates an item within a sequence of page numbers, just as a catalog record locates an item within a sequence of call numbers.

To treat catalog records and tables of contents entries as data points, instead of just location pointers, dislocates these familiar forms from their usual function, revealing expanded possibilities for analysis. Gitelman and Jackson point out that "data are aggregative," suggesting that the accumulation and organization of individual data points build meaning.[16] In the sections that follow, I transform the ordered sequence of a set of catalog records or the entries in a table of contents into relational analytic structures that enable the exploration of connections that the form of the list obscures.

The Field of Victorian Poetry in Library Records

Library collections offer selective virtual representations of the cultural record: even national repository libraries, such as the British Library and the Library of Congress, cannot be entirely comprehensive in their collection and preservation of printed material. Although the questions that motivate scholarship today are frequently driven by different aesthetic, political, and historical assumptions than they were even a few decades ago, the resources available to us are shaped by decisions made by previous generations of book collectors and librarians. Even as the Google Books Library Project and other large-scale digitization efforts provide us with an increasingly detailed and data-rich virtual Victorian reality, it is worth remembering that these digital surrogates also result from accumulated past decisions about the value of particular books. This incomplete and virtual representation of the Victorian period is gradually becoming the only one available, as the processes of decay and entropy privilege those Victorian items that have been preserved by libraries.

Cataloging practices have become increasingly standardized over time, but for items cataloged before the mid-twentieth century, library records can vary even for identical copies of the same book. The existence of multiple records that share the same basic bibliographical metadata (such as author, title, publisher, and date) in the National Union Catalog and in WorldCat may reflect some distinguishing features of a particular copy, a different interpretation of the bibliographic evidence, or a lack of knowledge about copies held in other libraries.[17] For some analytic purposes, such records could be combined without losing important information; for others, they need to be kept separate.[18] In the following analyses of the virtual knowledge of Victorian publishing provided by library catalog records, I have not attempted to collapse multiple bibliographically similar records, for both conceptual and practical reasons. At the practical level, extensive comparison of the physical books in multiple libraries corresponding to similar, but distinct, catalog records would be required to estimate the significance of those distinctions and the prevalence of erroneous or insignificant duplication of records. More importantly, at the conceptual level, I am analyzing the cultural field as it is instantiated in the digitally accessible records of published books of poetry held today in research libraries. This is a virtual field, a space of information that is constituted by overlap and duplication as much as it is shaped by omission.

Creating the data sets discussed in this essay required transforming catalog records designed to locate specific items within material and conceptual hierarchies into a relational format that facilitates the exploration of connections among items. Because each decision made in selecting and refining data affects what can be explored and understood with that data, I document those selection processes in order to emphasize the constructed nature of the data and of the visualizations used to explore it. To create this data, I first constructed a series of queries in CQL (Contextual Query Language) with stemmed keyword searching and used the SRU (Search/Retrieve via URL) feature of the OCLC WorldCat Search API (Application Programming Interface) to download records in XML (Extensible Markup Language) format. I then extracted data from specified MARC (Machine Readable Cataloging) fields.

The creation of bibliographic metadata during the cataloging process locates each item within the library's conceptual hierarchy and on its physical shelves, which constrains the ways in which the item can be found by future users. In libraries following the Library of Congress system, for example, most books of poetry are cataloged under author names, which are sequenced under national and chronological categories. With the name of an author, or the title of a book, it is easy to find the relevant item, particularly because that information is often explicitly recorded in the material book itself. But other features of literary texts, such as genres and forms, are only partially evident in the bibliographic metadata and are not typically searchable from standard catalog interfaces. (Subject headings exist for genres, forms, and other descriptive features of literature, but the headings point to works about these categories, rather than to works belonging to these categories.) To find records for books of poetry, I developed a keyword search strategy that included both general and specific Victorian terms for poetry, including, *verse*, *poem*, *ode*, *elegy*, *sonnet*, and *lyric* (and their related plural forms). Keyword searches return items with the keyword present in any fields of the cataloging record, including volume title, descriptions, notes, or subject headings. This approach returns a large number of items, including many not listed in standard bibliographies of nineteenth-century poetry. Yet these results should still be considered partial, rather than comprehensive, as there are undoubtedly some records for books of poetry that lack the keywords I used.

To further refine the returned results, I also specified the item type as "book," restricted the language to English, and limited results to selected years between 1860 and 1875. Despite eliminating key American cities from the query results, the returned data contained a large number of non-British items that had to be removed during data preparation. Other data preparation tasks included standardizing references to publisher names (e.g., combining *E. Moxon*, *Edward Moxon*, and *Moxon* under one name), removing extraneous punctuation, and adjusting records with nonstandard fields.

Examining this data reveals the strong dominance of London as the major center of British poetry publishing in the midnineteenth century, as shown for the year 1862 in Figure 1. Such long-tailed distributions are commonly found in data displaying preferential attachment, the process by which resources are concentrated within a system. Typical examples include the tendency of the wealthier members of a community to gain additional wealth or, in this case, for cities with many publishers to attract more publishers.[19] The horizontal bar graph in Figure 2 shows the relative importance of London as compared with other major publishing centers, such as Edinburgh and Cambridge.[20] Plotting these data points on a map, as in Figure 3, reveals the geographical dispersion of poetry's publication throughout England, Scotland, and Ireland in 1862. Exploring multiple visualizations of the same data opens new questions for research into the cultural and economic roles that such publishing played at midcentury and into the many publishers then active who are nearly unknown today. Restoring them to view alters the frame through which we perceive both the center and the periphery of this field.

To gain a bird's-eye view of the cultural field of Victorian poetry, I use network analysis and visualization to explore poet-publisher relationships as they were materially instantiated in published books. Network analysis examines the connections (visually depicted as edges, or lines) between entities (visually depicted as nodes, or points) in complex systems. This analysis examines the number, type, and pattern of edges and nodes in order to understand the relative importance of particular nodes, the existence of interconnected clusters, and the flow of information (or social status, traffic, capital, and so on) from one node to another. The force-directed algorithms used to generate the network visualizations in this essay model the structure of a network by applying spring and electrical forces of attraction and repulsion to edges and nodes. This facilitates understanding of the connections within the network. In particular, because visualization reveals the overall structure of a network, "it can allow users to see relationships, such as patterns and outliers, that would not be apparent through a metrics-based analysis alone."[21]

In the network visualization in Figure 4, each item record in the data set for 1862 is represented by an edge connecting a poet with a publisher.[22] The degree centrality of each node, or the total number of edges connected to it, is a useful measure of its local importance. The two nodes with the greatest degree centrality in this network represent the publisher and author as "not listed." In some cases, the material books described by these catalog records may be lacking indicators of publisher or author names; in other cases, that information may have been omitted from the record or entered in a nonstandard MARC field. Multiple edges connected to a named author node in this graph represent books that are closely related by shared authorship. However, while the multiple edges connected to the "author not listed" node might be closely related or might be quite distinct, that information is not present in the data set. Were additional author names to be discovered, some of the items currently categorized as "author not listed" might turn out to be authored by the same person, and others would not. This ambiguity is a defining characteristic of the virtual Victorian reality available from library catalogs.

Most publisher and poet nodes in this network for 1862 are found in pairs or triples, which are dispersed to the periphery of the graph because they are not connected to the rest of the network. The high number of poets represented by only one item record in a particular year is expected, given the typical rate at which nineteenth-century poets published volumes of verse. But this data also reveals that many Victorian publishers issued only one or two volumes of poetry in a given year, usually not enough for them to be especially recognized for poetry. Poetry was part of the general literary production of a much wider range of publishers than has typically been recognized. These are the publishers on the long tail of the distribution, those less connected in the field and also generally less known today. The asymmetrical distribution of resources in the cultural field, which Bourdieu believes to be homologous across different cultural domains, persists today in our virtual Victorian reality as the large and well-known Victorian publishers tend to be the ones that remain in our awareness. As Figure 5 suggests, several of the dominant publishers of books of poetry in the 1860s still exist today, and others will be quite familiar to readers knowledgeable about the period. This network reveals one view of the overall structure of the field of Victorian poetry publishing for the year 1862. Further research—comparing other books issued by each firm, the prices of these books, and additional factors that contributed to the social, economic, and cultural capital associated with different publishers at midcentury—would be required to conduct a fuller analysis of the cultural field and its effects.

In order to explore the production of poetry within major publishing centers during 1860–69, I created a second data set from the OCLC catalog records, limiting the place of publication to London, Oxford, Cambridge, Manchester, Leeds, Liverpool, Edinburgh, Glasgow, or Dublin. Again, it is important to note that the data collected by this method represents only what is available from digital records of books that are held in research library collections.[23] This data can thus only ever be a partial representation of the field of poetry book publishing. When visualized with the same force-directed algorithm, the data in Figure 6, for the decade 1860–69, reveals a similar graph structure to that of the year 1862: publishers with many connections to authors are at the center of the network, and publisher-author pairs or triples are dispersed to the periphery.[24] This visualization clearly conveys the importance of high-degree publishers and authors, as well as the distinction between them and those less well connected.

As in the 1862 data set, the two largest nodes in the network represent items with no publisher or no author listed. Highlighting the "publisher not listed" node in Figure 7 reveals a tight group of author nodes connected to it, as well as others arranged through the central space of this view of the field of poetry. These nodes represent authors for whom at least one item record has no publisher listed, but who are connected to named publishers in other records. Figure 8 highlights the "author not listed" node, which can occur for anonymously published works or for those records that lack author information for other reasons. The thickness of the edges in this graph reflects the edge weight, or the number of times the same pair of nodes are connected. The thick edge connecting the "no publisher" and "no author" nodes indicates 1,967 items lacking both these pieces of metadata. Missing metadata is a constitutive feature of the Victorian field that these records comprise, and it directs attention toward or away from particular positions in the field.

Because this network consists of data from multiple years, it reveals several distinct groups of authors connected to one publisher, whereas other authors are placed throughout the central space of the field because they are connected to multiple publishers. Figure 9 highlights the nodes for authors published by Macmillan, who include Christina Rossetti, Dante Gabriel Rossetti, William Allingham, Francis Turner Palgrave, Coventry Patmore, Richard Chenevix Trench, Dinah Mulock Craik, and Augusta Webster, as well as lesser-known figures, such as Georgiana Chatterton, Fanny Wheeler Hart, and Edmund Sandars. This data could enable scholars to explore the connections among works published by the same firm. Such research might examine the social or familial links among poets (as in the Rossetti circle), the aesthetic or topical features of the poetry (as in the volumes put out by religious publishers), or the bibliographic features of the books as material objects (such as paper, typeface, and bindings). All of these kinds of connections contributed to the structure and perception of the cultural field of Victorian poetry.

Node degree in this graph represents the number of catalog records associated with an author. Due to duplication in the original data, these numbers should be understood as reflecting the structures of that data (that is, the library catalog records) rather than a strict bibliographic measure of books or editions published. Of the 2,520 named authors on this graph, 80 percent have a node degree of 1, meaning that they are represented by only one item in the data set (see Figure 10). Authors with degrees of 1 or 2 (totaling 2,345) make up 93 percent of the data set. The higher-degree authors (ranging from degrees 3 to 31) include nineteenth-century poets, such as Burns, Byron, Longfellow, Scott, Wordsworth, Hood, Keble, and Tennyson, along with those from earlier periods, including Shakespeare, Milton, Cowper, and Thomson, as well as poets in translation, such as Horace, Virgil, and Goethe (Figure 11). Many of the canonical poets with whom we are most familiar are positioned near the center of the graph, which represents the dominant area of the field, as they are well connected with high-degree publishers. But it is important to recognize that proportionally they constitute only 7 percent of the total number of authors represented in this data.

Examining the distribution of node degree for the named publisher nodes (shown in Figure 12) reveals that 74 percent of the named publishers in this network have a node degree of 1, and publishers with degrees of 1 or 2 make up 84 percent. The node degree distribution is different for publishers than for authors, of course, in that there are also a small number of well-connected publishers with node degrees between 50 and 116, including Longman, Bell and Daldy, Macmillan, Routledge, Simpkin, Marshall, Moxon, and Trübner. As familiar as scholars might be with some of the social, economic, and cultural values associated with these well-known publishers, there are hundreds of others whose presence in the cultural field of Victorian poetry has not yet been fully considered or understood. Network analysis typically focuses attention on the high-degree, well-connected nodes. This path of analysis will continue to provide insight into the structure of the cultural field and its dominant positions. But it should also remind us of the poets and publishers at the periphery of that field. As Bourdieu notes, "every position, even the dominant one, depends for its very existence, and for the determinations it imposes on its occupants, on the other positions constituting the field."[25] We can better understand the variety of positions in the cultural field by analyzing publication information at a larger scale.

THE FIELD OF VICTORIAN POETRY ANTHOLOGIES

Anthologies of literature, especially those that are used in classrooms, both represent and contribute to the structure of the literary field, since the contents of a particular anthology metonymically stand in for "the nineteenth century" or "the Victorians." Although editors and instructors know such a Victorian field to be virtual in its incompleteness (a fact made evident in the digital supplements now frequently offered online by textbook publishers), the anthology becomes a tangible version of the field as it guides the reader's experience. In a pedagogical context, the selections an instructor makes from a particular anthology form an additional subfield, with explicit or implicit valuation being assigned to the works on the syllabus as compared with those in the other pages of the anthology. In the sections that follow, I treat the anthology table of contents as a storehouse of cultural data, dissociating it from its usual functional purposes (which include classifying the volume's content text with the metadata of author names and document titles, directing a reader to a particular location in the material volume, or outlining the scope of a field of study, whether historically, formally, or thematically constituted).

For this analysis, I have selected five anthologies of Victorian poetry published within an eight-year span (1997–2004) and still in print as of this writing in 2014. I have chosen to focus here on specialist anthologies, rather than on the surveys of British literature published by Norton, Longmans, and Broadview (which also include some Victorian poetry), or on cross-genre Victorian anthologies like Herbert Tucker and Dorothy Mermin's *Victorian Literature: 1830–1900* (published 2001). Each of these collections, which are aimed at the classroom market and possibly also at specialized general readers, presents itself as an instantiation of the field of Victorian poetry:

- *The Penguin Book of Victorian Verse*, ed. Daniel Karlin (Penguin, 1997)
- *The Broadview Anthology of Victorian Poetry and Poetic Theory*, ed. Thomas Collins and Vivienne Rundle (Broadview Press, 1999)
- *The Broadview Anthology of Victorian Poetry and Poetic Theory, Concise Edition*, ed. Thomas Collins and Vivienne Rundle (Broadview Press, 2000)
- *Victorian Poetry*, ed. Valentine Cunningham and Duncan Wu (Blackwell Essential Literature series, 2002)
- *Victorian Poetry: An Annotated Anthology*, ed. Francis O'Gorman (Blackwell Annotated Anthologies series, 2004)

In the sections that follow, I will refer to these as the Penguin, Broadview Full, Broadview Concise, Blackwell Essential, and Blackwell Annotated anthologies.[26]

Literature anthologies contribute to the process of canon formation by representing literary history through the careers of selected writers, by reproducing frequently selected texts, and by intervening in those established practices with new selections (which over time may persist and gain canonical or classroom value, or may fade from view if not included in subsequent anthologies). All of these actions occur within and contribute to a literary field that delimits the range of possible authors and texts and assigns their cultural value. Anthologies offer a good small-scale example of Bourdieu's point that a cultural field is simultaneously a space of possibilities, a "force-field acting on all those who enter it . . . in a differential manner according to the position they occupy there," and "a field of competitive struggles," the outcomes of which continually alter the structure of the field itself.[27] The tables of contents in these five anthologies thus provide data about how the field of Victorian poetry is currently constructed, not simply for their readers, but also for future collections that may seek to replicate or overturn the dominant selections and values of the field.

The table of contents is, in Gérard Genette's terms, an element of the paratext: "a zone between text and off-text, a zone not only of transition but also of *transaction*."[28] The table of contents affords the reader's transactions with the text by describing the book's textual content as filtered through the material format of the codex, in which the folded leaves of paper are conventionally represented with an ordered series of numbers indicating pages.[29] The table of contents thus translates the spatial, material location of information within the codex into a numeric sequence or list. Such locations could be represented in alternative terms, such as a measurement of the thickness of the book or a reduced visual representation of what the page in question looks like, but even in digital reading environments the convention of the table of contents as a list remains strong. The tables of contents in the five anthologies I examine here also reveal the ideological function of the author as an organizing structure: each uses author names, including dates of birth and death, as headings under which individual titles for poems are listed. This is a prevalent feature of literature anthologies designed for classroom use and exists in many general-purpose anthologies as well.[30] Each of these anthologies constructs its version of Victorian poetry with a different number of poets. As Figure 13 reveals, the Penguin anthology contains more than twice the number of the Broadview Full, which contains more than twice the number of the Blackwell Annotated or the Broadview Concise. These last two volumes more than double the number of poets in the Blackwell Essential. Such differences significantly impact not only the variety of each collection's texts, but also the model of literary history each reflects and constructs. Comparing the number of male and female poets included in each anthology is facilitated by visualizing the relative percentages of each, as in Figure 14. Women poets make up almost a third of the authors in each collection, with the exception of the Broadview Concise, in which women poets constitute 44 percent of the total number. (Of course, such ratios only account for a poet's inclusion in the anthology, not for the number of poems or pages by which he or she is represented.)

Together, these five anthologies contain works by 153 poets, who can be usefully sorted into five groups: those included in all five collections, those in four, and so forth. These frequency groups offer a simple metric for canonicity in the anthology field, with the 63 poets in groups 2–5 making up the general Victorian poetry canon. Figure 15 offers insight into how clearly aligned each anthology's selection of poets is with the larger field. We see nine poets—Emily Brontë, Elizabeth Barrett Browning, Robert Browning, Arthur Hugh Clough, Thomas Hardy, Gerard Manley Hopkins, Christina Rossetti, Dante Gabriel Rossetti, and Alfred Tennyson—appearing in all five anthologies. The Penguin collection, which features the greatest total number of poets, is also the collection with the greatest number of distinctive choices not found in the other volumes. Bringing the author lists from each anthology together for analysis illuminates the contours and conventions of the field; this is shown in the radial visualization in Figure 16, in which each anthology node is connected to its poet nodes, which are ordered by frequency group.

These tables of contents present each poet's lifespan as important information that also dictates the ordering of materials in each anthology. Comparing the birth, death, and midlife dates for the 63 poets included in two or more volumes reveals a fairly even distribution of dates of birth between 1800 and 1850, except for a peak number of births between 1820 and 1825, and a peak number of deaths after 1880 (Figure 17). Visualizing each poet's lifespan individually, as in Figure 18, provides another way of exploring this chronological distribution in conjunction with gender. This visualization reveals, for example, that most of the women poets were born before 1840 and that the short lifespans of poets like Ernest Dowson and Amy Levy are fairly exceptional in this field. Given the long chronological period encompassed by these anthologies, comparing poets' lifespans allows one to discover subgroups and develop metrics for generational comparisons. For example, marking the midway point in each poet's life on the individual timelines (Figure 18) highlights generational distinctions among poets who may have lived and published at the same time. Additional analysis could add dates of publication to the timelines to further complicate this biographical framework. (When dates of composition or publication are included in these anthologies, they are only indicated with each printed text, not in the table of contents.)

Of course, the field of Victorian poetry as it is constituted in these anthologies consists not simply of author names, but also of specific poems for each author. These selections can vary considerably from anthology to anthology, even for the nine poets who are present in all five. To explore how each poet's works structure the field, I again use network analysis, creating a network for each poet in which the nodes represent poems, and the edges between them are weighted by the number of times that textual pair occurs within an anthology in the data set (thereby producing weights ranging from one to five). For example, Christina Rossetti's "Goblin Market" and "A Birthday" occur together in four anthologies, so the weight of the edge between those two nodes is set at four. Figure 19 shows that Rossetti's poetry falls into three distinct communities in this field: a group of 13 poems that have strong ties to one another because they are printed together in several anthologies, a large group of weakly connected poems in the lower half of the graph, and a smaller group at the top right (reflecting the fact that the Penguin and Blackwell Annotated anthologies print selections that do not appear in other collections).[31] With nodes colored according to the frequency with which they appear in the data set, this visualization allows us to explore the relationship between frequency and co-occurrence of texts in anthologies. As shown in Figure 20, the nodes for Rossetti's "Winter: My Secret," "A Christmas Carol," and "Remember" have a frequency of two but are located on the graph among the nodes for the Penguin collection, as they are included in that anthology and thus co-occur with its other poems. The six nodes with a frequency of two at the upper left edge of the network are all included only in the Broadview Concise and Broadview Full anthologies. Those nodes are placed closer to the central, most frequent texts (like "Goblin Market" and "A Birthday") because the contents of these two Broadview collections are more closely connected to that central group. These visualizations reveal the poems that most typically represent Rossetti's poetry in the field of the anthologies.

The overall structure of the network of poems by Gerard Manley Hopkins in Figure 21 reveals less divergence in the selections, as those poems with lower frequencies co-occur with the most canonical texts.[32] All of the nodes in this graph are connected in some way to the central, high-frequency nodes, which include two poems present in all anthologies ("The Windhover" and "As kingfishers catch fire") and four that are included in four anthologies (Figure 22). Because poems with a frequency of one occur in all the anthologies except the Broadview Concise, those low-frequency nodes are more evenly distributed around the edges of this graph than they were in the previous one. Even at a glance, such network visualizations suggest that the representation of Hopkins's poetry in the field of the anthologies is more consistent than that of Rossetti.

Extending this kind of analysis allows us to discover which poems by different authors tend to co-occur in the same anthology, which reveals the textual structure of this literary field. (*Co-occurrence* in this analysis is defined simply as being printed in the same collection, not by any additional within-volume measures, since texts in these volumes are all organized by author birthdate.) Figure 23 visualizes the co-occurrence of poems by the nine authors who appear in all five of these anthologies.[33] Each node in the network represents a poem, and the weight of the edge between it and another poem is the number of times those two texts occur in an anthology together (thus there are 26,620 edges for 309 nodes). Again, the most connected nodes are placed in the center of the graph, with less connected nodes at the edges. This is a highly interconnected graph, since even a poem like Hopkins's "The Caged Skylark," which is only printed in one anthology (Broadview Full), will be connected to every other poem in that anthology. Understanding the deeper structures of the anthology field requires more than simply counting authors or even cataloging the frequency with which specific texts are reprinted. This co-occurrence network begins to reveal the contours of the field as it is experienced by readers of a particular anthology, who encounter each poem in the network of the other selections in that volume.

Applying the Louvain community detection algorithm reveals the five communities of interconnected nodes labeled A–E and color-coded on the graph in Figure 23.[34] Degree centrality of the nodes within most of these communities varies because the frequency and distribution of the nodes over the five anthologies varies. For example, nodes in community C (colored green) are located in three areas of the graph in Figure 23, forming three smaller subgroups. Whether the editors of these collections intended it or not, the communities displayed in Figure 23 inevitably influence the perceptions and possible actions of all who enter the anthologies' literary field. At the simplest practical level, a student asked to compare three poems for a paper will likely select them from the texts in the course anthology. At a more subtle level, it is likely that poems in the same community share aesthetic, ideological, or historical features that have not yet been fully explored. For example, at the center of the graph, a group of seven nodes in community C includes Christina Rossetti's "Song (When I am dead, my dearest)," "In an Artist's Studio," and "A Birthday," clustered together with Elizabeth Barrett Browning's "The Runaway Slave at Pilgrim's Point," Tennyson's "The Charge of the Light Brigade," Hardy's "In Tenebris," and Hopkins's "No worst, there is none" (Figure 24). Yet other poems by Rossetti, Tennyson, Hardy, and Hopkins belong to community B (colored lavender). These include "Goblin Market," "Crossing the Bar," "The Darkling Thrush," and "The Windhover." Exploring this data through network visualization raises questions for further analysis and interpretation, which might involve close reading these groups of poems to discover formal, aesthetic, or thematic connections. A complementary algorithmic approach might identify multiple categorical features of these poems (such as rhyme scheme or diction) and apply correspondence analysis to understand the structure of distinction at work in this field.[35]

As I've suggested here, digital reading combines the computational analysis of quantitative data with the humanist awareness of the subjective processes at work in creating that data. Reimagining functional tools of scholarship, like the library catalog record and the table of contents, as stores of data opens up new paths for exploratory analysis. Data visualization can reveal complex relationships in large sets of data and help us better understand the structures of the cultural field of Victorian poetry. As we reconstruct the historical contours of that field and excavate the contemporary fields that shape our ideas of Victorian poetry, we add to the knowledge that generates our virtual Victorians.

NOTES

1. Katherine Bode, *Reading by Numbers: Recalibrating the Literary Field* (London: Anthem Press, 2012), 7–26.
2. Natalie M. Houston, "Toward a Computational Analysis of Victorian Poetics," *Victorian Studies* 56, no. 3 (2014): 498–510.
3. Pierre Bourdieu, *The Rules of Art: Genesis and Structure of the Literary Field*, trans. Susan Emanuel (Stanford: Stanford University Press, 1996), 231.
4. See Gisèle Shapiro, "Autonomy Revisited: The Question of Mediations and Its Methodological Implications," *Paragraph* 35, no. 1 (2012): 37. doi: 10.3366/para.2012.004037.
5. Peter D. McDonald, *British Literary Culture and Publishing Practice, 1880–1914* (Cambridge: Cambridge UP, 1997), 12–13.

6. Pierre Bourdieu, "The Field of Cultural Production, or: The Economic World Reversed," in *The Field of Cultural Production: Essays on Art and Literature*, ed. Randal Johnson (New York: Columbia UP, 1993), 31.
7. Johanna Drucker, "Humanities Approaches to Graphical Display," *Digital Humanities Quarterly* 5, no. 1 (2011): par. 3. Accessed February 24, 2014,http://www.digitalhumanities.org/dhq/vol/5/1/000091/000091.html.
8. Lisa Gitelman and Virginia Jackson, "Introduction," in "*Raw Data" Is an Oxymoron*, ed. Lisa Gitelman (Cambridge: The MIT Press, 2013), 3.
9. Hans Robert Jauss, *Towards an Aesthetic of Reception* (Minneapolis: University of Minnesota Press, 1982), 19.
10. John W. Tukey, *Exploratory Data Analysis* (Reading, MA: Addison-Wesley, 1977), v, 2–3.
11. Frederick W. Gibbs and Daniel J. Cohen, "A Conversation with Data: Prospecting Victorian Words and Ideas," *Victorian Studies* 54, no. 1 (Autumn 2011): 70.
12. Tukey, *Exploratory Data Analysis* 56.
13. The data visualizations for this essay were created using Gephi (https://gephi.github.io/), R (http://www.r-project.org/), and Excel (http://office.microsoft.com/en-us/excel/). They are available in the *Virtual Victorians* online annex.
14. See Michel Foucault, "What Is an Author?" trans. Donald F. Bouchard and Sherry Simon, in *Language, Counter-Memory, Practice*, ed. Donald F. Bouchard, (Ithaca, NY: Cornell UP, 1977), 124–27.
15. John Guillory, *Cultural Capital: The Problem of Literary Canon Formation* (Chicago : University of Chicago Press, 1993), 33.
16. Gitelman and Jackson, "Introduction" 8. The ambiguity in current English usage of the plural noun *data*, which is used with both singular and plural verbs, reflects its etymological and cultural history. See Daniel Rosenberg, "Data before the Fact," in "*Raw Data" Is an Oxymoron*, ed. Lisa Gitelman (Cambridge: The MIT Press, 2013), 15–40.
17. WorldCat is the online union catalog created by the Online Computer Library Center (OCLC), which contains records from over 70,000 member libraries. See http://oclc.org.
18. The entity-relationship model of the 1998 Functional Requirements for Bibliographic Records (FRBR) constructs an ontology of work, expression, manifestation, and item that can be used to merge the display of records to meet the needs of some users. Such collapsing of distinctions may not be appropriate for all forms of research. See IFLA Study Group on the Functional Requirements of Bibliographic Records, *Functional Requirements of Bibliographic Records: final report*, IFLA Series on Bibliographic Control 19 (Munich: K.G. Saur Verlag, 1998), http://www.ifla.org/publications/functional-requirements-for-bibliographic-records.
19. Long-tailed distributions, including power law and Pareto's law distributions, can be found in a wide variety of data, such as word occurrences,

meteorological phenomena, social network behavior, and financial systems.

20. All multicolored graphs have been produced following the guidelines in Masataka Okabe and Kei Ito, "Color Universal Design (CUD): How to Make Figures and Presentations That Are Friendly to Colorblind People," J*Fly Data Depository for *Drosophila* researchers, http://jfly.iam.u-tokyo.ac.jp/color/. As of February 2015 this website is unavailable. Okabe and Ito's colorblind safe palette is reproduced in Brian Connelly, "Creating Colorblind-Friendly Figures," *Brian Connelly,* October 16, 2013. Accessed September 21, 2014, http://bconnelly.net/2013/10/creating-colorblind-friendly-figures/.
21. Helen Gibson, Joe Faith, and Paul Vickers, "A Survey of Two-Dimensional Graph Layout Techniques for Information Visualisation," *Information Visualization* 12, no. 3–4 (2012): 325. doi: 10.1177/1473871612455749.
22. This visualization was created in Gephi with Yifan Hu's force-directed algorithm, with an optimal distance of 50 and initial step size of five. See Yifan Hu, "Efficient, High-Quality Force-Directed Graph Drawing," *Mathematica Journal* 10, no. 1 (2005): 37–71.
23. Not all records from the earlier *National Union Catalog Pre-1956 Imprints* series are yet available through OCLC. See Christine DeZelar-Tiedman, "The Proportion of NUC Pre-56 Titles Represented in the RLIN and OCLC Databases Compared: A Follow-Up to the Beall/Kafadar Study," *College & Research Libraries* 69, no. 5 (2008): 401–06.
24. This visualization was created in Gephi with Yifan Hu's force-directed algorithm, with an optimal distance of 50 and initial step size of five.
25. Bourdieu, "The Field of Cultural Production" 29–30.
26. As part of the Blackwell "Essential Literature" series, the Cunningham and Wu anthology contains a selection of texts from the 1100-page *The Victorians: An Anthology of Poetry and Poetics* (published in 2002, also edited by Cunningham, and now out of print).
27. Bourdieu, *Rules of Art* 232.
28. Gérard Genette, *Paratexts: Thresholds of Interpretation*, trans. Jane E. Lewin, Literature, Culture, Theory 20 (Cambridge: Cambridge UP, 1997), 2.
29. Missing or unordered pages are generally considered exceptional and worthy of attention in specialized cases made evident to us through descriptive bibliography, itself a technology for transforming the material codex book into sets of data. See Fredson Bowers, *Principles of Bibliographic Description* (Princeton: Princeton UP, 1949).
30. Thematic or strictly chronological organization is more rare, as seen in Francis Turner Palgrave, ed., *The Golden Treasury of the Best Songs and Lyrical Poems in the English Language* (Cambridge: Macmillan, 1861) or Jerome J. McGann, ed., *The New Oxford Book of Romantic Period Verse* (Oxford: Oxford UP, 1993).

31. This visualization was created in Gephi with the Fruchterman-Reingold force-directed algorithm, which works well for networks with fewer than 50 nodes. See Thomas M. J. Fruchterman and Edward M. Reingold, "Graph Drawing by Force-Directed Placement," *Software: Practice and Experience* 21, no. 11 (1991): 1129–64.
32. This visualization was created in Gephi with the Fruchterman-Reingold force-directed algorithm.
33. This graph was created in Gephi using the Force Atlas force-directed algorithm, which is optimized for community detection. See Mathieu Bastian, Sebastien Heymann, and Mathieu Jacomy, "Gephi: An Open Source Software for Exploring and Manipulating Networks," *ICWSM* 8 (2009): 361–62.
34. See Vincent D. Blondel, Jean-Loup Guillaume, Renaud Lambiotte, and Etienne Lefebvre, "Fast Unfolding of Communities in Large Networks," *Journal of Statistical Mechanics: Theory and Experiment* 2008, no. 10 (2008): P10008. doi: 10.1088/1742-5468/2008/10.
35. Bourdieu's analyses of the structures of the literary field in nineteenth-century France, for example, use multiple correspondence analysis, which distributes categorical features in Euclidean space to illuminate the underlying structures in complex data. See Wouter De Nooy, "Fields and Networks: Correspondence Analysis and Social Network Analysis in the Framework of Field Theory," *Poetics* 31, no. 5 (2003): 305–27.

Part II

Virtual Imaginings

Chapter 7

Virtual Victorian Poetry

*Alison Chapman**

N.B. Additional images associated with this chapter are housed in the digital annex at www.virtualvictorians.org.

In 1895, reviewing his attempt over 20 years earlier to provide an account of Victorian poetry—to define, categorize, and canonize the genre—E. C. Stedman commented that "even the adjective 'Victorian' was unfamiliar, if it had been employed at all."[1] But the adjective "Victorian" still required explication in his *Victorian Anthology 1837–1895*, as well as in his critical book *Victorian Poets*.[2] In fact, defining the field of Victorian poetry was a question many critics and poets reflected upon at the end of the century, especially in essays, reviews, and poems published in periodicals (by which I mean all forms of serial ephemeral print). This chapter addresses attempts to categorize poetry's value at the end of the nineteenth century through the cultural work of poems published in periodicals. At a time when the body of poetry from the Victorian era was under scrutiny, the status of poetry within serial print became a particular marker of modernity: it signified immersion in a virtual world, but also reflected that virtual world's limitations.

The prominent editor, critic, and poet Oscar Wilde determined that women poets were particularly important at the century's end

* I would like to thank the graduate students in my Victorian poetry seminar in the fall of 2013 for lively discussions on periodical poetry, Olivia Ferguson for her research assistance, the benefactor who gave the monthly issues of *Woman's World* to the University of Victoria Special Collections, and the university librarians for invaluable support.

not just for their impressive artistry, but also "for the light they throw upon the spirit of modern culture."[3] A poet featured prominently in Wilde's monthly magazine *Woman's World* was Mary C. Gillington (1861–1936), one of the most prolific (and arguably one of the least known) writers of this period.[4] In 1889, the magazine published a set of pastoral love poems by Gillington. Her 12-part sequence suggests that, during the late Victorian period, at a time when sustained critical and editorial attempts were made to define the era's poetry as "Victorian," poems in ephemeral print respond to similar questions about the place and value of poetry. Gillington's highly stylized poems follow the rural calendar to narrate the shepherd-speaker's courtship of his beloved milkmaid, Amabel, from her first greeting at the beginning of the year to the requited love of Christmas Eve. Each poem is set in antiquated type with a framed pastoral illustration that includes a decorative banner heading for the month. The 12 periodical contributions comprise serial poetry: a cycle of monthly installments that depicts courtship in a conventional narrative arc. Using Gillington's poems as a test case, this paper explores poetry's engagement with the conception of time embodied in periodical print culture, the hypertemporality of a virtual world that is date-stamped and yet cyclical.

The term *virtual* as it is now primarily understood, in terms of computerized or digitized virtual reality (OED A.9), was of course unavailable in the Victorian era. Nonetheless, the cognate denotation of *virtual* as a supposed, imagined, or notional world was very much alive in the period. The other closely related Victorian denotation of the adjective *virtual* is "that may be so called for practical purposes, although not according to strict definition; very near, almost absolute" (OED A.4.b). Yet digitally produced virtual reality has a historical precursor in nineteenth-century representational axioms. Recently, scholars have deployed the word *virtual* to explicate the era's representational strategies of readerly immersion in a narrative and visual world.[5] The most extended discussion is offered by Alison Byerly, who argues that the development of literary tourism has deep associations with literary realism. She contends that narrative immersion in the realist novel simulates a similar rhetorical strategy in virtual travel that transports readers into the world of the narrative, so that "the virtual becomes the (fictional) real."[6] The Victorian novelistic virtual world, however, is "an effort at similitude that is ultimately unrealizable. The word *virtual* evokes an imaginative experience that is not fully realized, but aspirational. It is a stage of *almost* being, of tending toward realization."[7] Byerly's use of the term *virtual* invokes the OED meaning of a world that is both notional and simulated, but that ultimately encodes its own fictiveness.

The virtual world of print has particular connotations for periodical culture. Hilary Fraser, Stephanie Green, and Judith Johnston, for instance, describe the nineteenth-century periodical press as "a network of virtual, as well as actual, communities, enabling marginal interventions in mainstream debates, providing evidence of a society in transition."[8] James Mussell argues that the experience of reading a Victorian periodical title was made cohesive by the repetition of that title's formal features (including types of content, such as correspondence columns, reviews, and editorials, as well as stylistic features, such as masthead, layout, and paper quality), even if those features changed over time, because the repetition created "virtual forms" that were familiar to the reader.[9] Rhythms of repetition and variation in a periodical's formal features meant that single issues were, Mussell claims, virtually connected to the larger run, even if they could not figure metonymically for the entire series.[10] In effect, then, for Mussell, periodicals and newspapers are "virtual entities that are created in the spaces between individual issues."[11] Serial, ephemeral, popular periodicals are thus arguably the most virtual of Victorian print media, contingent on materiality and immateriality.

Variability and ephemerality were precisely the features of periodicity that some late-century literary titles explicitly addressed. A prominent example is *The Yellow Book* (1894–97), whose bright yellow cover, five-shilling price tag, and quarterly issue all indicated its ambition to be more like a book than a disposable periodical. In contrast, *Woman's World* (November 1887–October 1890, with Wilde's editorship running November 1887–October 1889) exploited the distinction between magazine issue and collated annual volume, emphasizing the disposable ephemerality of the monthly part with its thin paper, delicate pink cover, and generous advertisement sheets, and the heftiness of the book volume with its substantial cover and better-quality paper. Under Wilde's editorship, *Woman's World* offered the serial magazine as both a throwaway and a keepsake. Although this simultaneous investment in synchronic and diachronic time is most obvious in the contrast between the monthly issues and the collated annual volume, the magazine's contents also mirror this temporal doubling as they juxtapose essays on the reader's immediate context with essays on women in history.[12] For *Woman's World*, such issues of materiality and immateriality, original and copy, ephemerality and permanence, play out through the creation of a virtual, simulated world, an immersive and richly illustrated feminine space crowded with beautiful surfaces and gorgeous things.[13]

Critical to this virtual space is the regular inclusion of 1 or 2 poems in each issue that signified the commodification of the magazine by means of their decorative, aesthetic value-laden layout. In the period of Wilde's editorship, the magazine published 39 poems, all but 5 by women; 24 of the women's poems were illustrated (with either a picture or an ornamental embellishment). But in addition to being a signifier for a beautiful thing, an object to consume, each poem engages with the key issues of the magazine—issues that were also part of the discourse about the definition and status of poetry at the end of the century. Gillington's illustrated contributions, for instance, form a serial seasonal poem that illuminates the relationship between poetry and simulation. This example suggests not only that poetry is at the center of the construction of virtuality in the Victorian era, but also that understanding how that process occurred reveals we might only know Victorian poetry virtually, as a notional category.

Defining Victorian Poetry

M. C. Gillington may be one of the most prolific writers of her era, although she is little known today. Three prominent anthologies edited by Elizabeth A. Sharp, designed to represent and categorize women's poetry at the end of the century, feature Gillington, as well as her sister Alice. Both poets appear in Sharp's 1887 collections, *Women's Voices*, where they are placed at the very end of the book, and *Sea-Music*, where M. C. Gillington has seven entries.[14] Both sisters also appear in what is arguably the most foundational anthology to define the canon of women's poetry at the end of the century, Sharp's 1890 *Women Poets of the Victorian Era*. This chronologically arranged volume concludes with two poems from each sister, as if to signal that the Gillingtons are the rising stars of modern poetry.[15] These three anthologies were part of a wider movement to conceptualize the era's poetry at the end of the century and, indeed, to define what "Victorian poetry" meant as a classification for women poets, as well as for poetry in general. Other prominent books that staked a claim to the poetic canon of the era include Edmund H. Garrett and Edmund Gosse's *Victorian Songs*, Hugh Walker's *The Greater Victorian Poets*, and Stedman's *Victorian Poets* and *Victorian Anthology* (which included poems by both Gillington sisters).[16] As Leah Price observes, late-nineteenth-century editors attempted "to re-order a threateningly late and shapeless reading public" in the production of anthologies to excerpt and discipline their reading.[17] While Price's focus is on the novel, poetry anthologies also flooded the market toward the end of the century, with titles that indicate efforts to define specifically what Victorian poetry should look like.

Wilde was also engaged in defining Victorian poetry, and in particular the woman poet. In his December 1888 essay "English Poetesses," published in *Queen* when he was one year into his editorship of *Woman's World*, Wilde comments on the "really remarkable awakening of women's song that characterizes the latter half of our century in England. No country has ever had so many poetesses at once."[18] He attributes this female poetic renaissance to Elizabeth Barrett Browning and to women's education, but in a typical Wildean move he declares that there may be a surfeit of poets: "when one remembers that the Greeks had only nine muses, one is sometimes apt to fancy that we have too many." Nonetheless, he softens the blow by admitting immediately afterward that "the work done by women in the sphere of poetry is really of a very high standard of excellence." The solution (foreshadowing Gillington's future career) is that women poets should write prose, because "it is not that I like poetical prose, but I love the prose of poets." In his essay for *Queen*—ironically, as his competitor title, it was the exemplary upper-class woman's fashion magazine, "the glossy to end all glossies"[19]—Wilde positions himself as a sagacious commentator who shapes the traditions of, as well as the contemporary fashion for, women's poetry. As Kathryn Ledbetter argues, "women's poetry had a notable function in the mission of *Woman's World*," at a time when "the idea of a woman poet, rather than a conventional poetess, was relatively recent."[20] The assertion of the importance of women as poets was at the core of *Woman's World*, despite original intentions for the magazine not to include much poetry.[21] Notably, Gillington's cycle of poems began one month after the publication of Wilde's *Queen* essay.

In the 1880s and 1890s, Gillington was positioned as a prominent female poet who wrote in difficult genres popular with the Decadents, such as sonnets, villanelles, triolets, and rondels. Her poems appeared frequently in such periodicals as *Blackwood's Edinburgh Magazine*, *The Cornhill Magazine*, *The Spectator*, *Cassell's*, *Eclectic Magazine*, and *Littell's Living Age*. Like many Victorian poets, Gillington's early periodical publications led to a volume of poetry: *Poems* (1892), coauthored with her sister. Her appearance in *Woman's World* in 1889 was remarkable for a young writer; securing a poem a month for the entire year (although this ultimately played out rather differently, as addressed below) was a commitment never repeated under Wilde's editorship.

But what is even more extraordinary about Gillington's career is the shape it took after her 1892 *Poems*. Although she went on to publish more periodical poetry and more volumes of poems, such as *Twelve Original Christmas Carols* (1893), her book publishing diversified and intensified rapidly. Under the name "M. C. Gillington," her later married name "Mary Byron" (her husband was a descendent of Lord Byron's second cousin), and the pseudonym "Maurice Clare," Gillington became well known as an abridger of literary classics for children, a children's poet and prose writer, a song writer and composer (especially of music for schools), a religious author, a cookbook author, a travel writer, an editor of anthologies and novels, and the author of dozens of titles in the biographical series *Days with the Great Composers*, *Days with the Great Authors*, *Days with the Great Poets*, and *Days with Victorian Poets*. The British Library catalogue credits her with 185 titles. Richard Ellman sees allusions to Gillington's 1913 biography *A Day with William Shakespeare* (published under the pseudonym "Maurice Clare") in the "Scylla and Charybdis" episode in Joyce's *Ulysses* (1922)—an echo that Molly Hoff also finds in Woolf's *Mrs. Dalloway* (1925).[22] These allusions suggest an even deeper connection between modernist writers and Gillington's contributions to the *A Day with [. . .]* series: the exploration of the link between time and modernity, a link that Gillington's poems for *Woman's World* also investigate. In an interview with *The Musical Herald* in 1904, Gillington claims of her voluminous songs and musical settings alone that "I have lost all count of my work—there has been so much of it. . . . So no one can accuse me of 'loafing.'"[23] Gillington's publications in *Woman's World* indicate an author savvy about the print market and the necessity of swiftly adapting to changing literary climates. After all, it was during wartime that she would reinvent herself as a cookbook writer.[24] Gillington's status as a poet also suggests her sensitivity to wider print cultures and available readerships, as well as the limits of considering a canon of Victorian poetry based on books of poems rather than on the richness and diversity of nineteenth- and early-twentieth-century print. Gillington's 1892 *Poems* was issued in the same year as Michael Field's *Sight and Song*, Richard Le Gallienne's *English Poems*, Arthur Symons's *Silhouettes*, and Oscar Wilde's *Poems*, but to position her as only a Decadent poet, which is how she fleetingly appears in a chronology of fin-de-siècle poems, suggests that we don't yet know Victorian poetry very well—and we might only know Victorian poetry virtually.[25]

Simulated Feminine Space

Wilde's debut as a magazine editor occurred during a period fixated on defining Victorian poetry, and especially women's poetry. Under his editorship, essays on female poets were published alongside contemporary women's poems, in order to delineate his version of this literary category. Indeed, Wilde's investment in women writers is exemplified by his first issue, in November 1887; it features an essay by Eveline Portsmouth entitled "The Position of Women," immediately followed by Violet Fane's sonnet "Hazely Heath," an autumnal poem that, as Ledbetter argues, "appropriately commemorates and contextualizes the moment of publication."[26] In addition, as the inaugural poem of the magazine, Fane's sonnet establishes the title's commitment to complex poetic form and women's poetry. Along with a short story by Amy Levy and an essay on Oxford women's colleges "By a Member of One of Them," the first issue also includes Wilde's review of Sharp's *Women's Voices*, which (as we have seen) ends with poems by the Gillington sisters.[27] Wilde lists contemporary poets included by Sharp, many of whom would also feature in *Woman's World*: Christina Rossetti, Augusta Webster, Harriet Eleanor Hamilton King, A. Mary F. Robinson, Dinah Mulock Craik, Jean Ingelow, Emily Pfeiffer, May Probyn, Edith Nesbit, Rosa Mulholland, Katharine Tynan, and Lady Charlotte Eliot.[28] Wilde comments, after giving this list, that "on the whole, Mrs. Sharp's collection is very pleasant reading indeed, and the extracts given from the works of living poetesses are extremely remarkable." He adds that this is true because they are technically excellent, but also, as we have seen in the introduction, because of "the light they throw upon the spirit of modern culture."[29] For Wilde, women's poetry is fashionable, intrinsic to the conception of the poetry of the era, but is also a signifier of the notional, aspirational spirit of the age—or the virtual world that Byerly argues is embedded in literary realism. Women's contemporary poetry, for Wilde, offers virtual access to the immediate context, a simulation of the real, which for Wilde is modernity itself.

This review set the tone for future columns, which emphasized writing and especially poetry by women, such as the review of Constance Naden's *A Modern Apostle, and Other Poems*, Field's *Canute the Great*, Robinson's *Poems, Ballads, and a Garden Play*, Craik's collected poems, Nesbit's *Leaves of Life*, Caroline Fitzgerald's *Venetia Victrix*, and Graham R. Tomson's *The Bird-Bride*.[30] The advertisements for the monthly parts mix fashion (ladies' clothing, hats, corsets, electropathic belts, and dress fabric), literature (books and magazines, especially those issued by the magazine's publisher, Cassell's, such as "Indispensable Books for Ladies"), and other female-oriented products, such as Woodward's gripe water (meant for mothers to administer to children, with an endorsement by Emily Faithfull) and a regular tipped-in pamphlet advertising Pear's Soap. The first issue of the magazine under Wilde announced implicitly and explicitly its investment in women—their place in society, education, the professions, history, and literature—and thus in crafting a popular female readership who would consume such messages.

Woman's World, revamped by Wilde from its low-brow previous incarnation as *Lady's World*, which he termed "a very vulgar, trivial, and stupid production, with its silly gossip about silly people, and its social inanities,"[31] crafts itself as a hypertemporal female space. It is a virtual world filled with high-quality illustrations of the latest styles from Paris, alongside serious essays on history, politics, the Woman Question, and literature. When it was relaunched in November 1887, the magazine was significantly longer and more lavishly illustrated than its predecessor, covering a wide range of subjects that included fashion and protofeminist politics, and that specifically targeted middle-class female readers.[32] Setting himself apart from the overly "feminine" *Lady's World* and similar titles like *Queen* and *Lady's Pictorial*, Wilde describes his editorial role as transforming the magazine into "the recognized organ for the expression of women's opinions on all subjects of literature, art, and modern life"; it should "deal not merely with what women wear, but with what they think, and what they feel." In this way, the magazine would take a "wider range" and "higher standpoint."[33] The title *Woman's World* suggests that the world created by the magazine was singular but also representational (the world of woman), particular yet also a synecdoche for female readers' lives. *The Times* pronounced the magazine "written by women, for women and about women" (in fact it was not entirely written by women).[34] Interestingly, *The Spectator* praises the 1889 collated annual volume issue not for its aspirational, notional world, but for its indexical link to reality—"fashions of dress, past and present, naturally occupy a considerable space, for the 'world' is an actual, not an ideal one"[35]—thus mixing together what Wilde terms "the things and the symbols of things."[36] Wilde describes the gendered space of the magazine, in a letter to Helena Sickert, as "representative of the thought and culture of the women of this century."[37] The pages of *Woman's World*, under Wilde's editorship, fashioned a virtual world for the reader's consumption, a simulacrum of middle-class fashionable life. The interests of the "new woman" implied by the title (as Laurel Brake notes) were to include protofeminist politics. Yet *Woman's World* was also a supremely female, aestheticized publication, one in which consumption of gorgeous objects was not counter to intellectual life, for ideas were also treated as things (as Stephanie Green argues). This is a notional gendered world that is stylized, decorative, and full of beautiful items: "gloves, fans, boxes, bottles, ornaments, hats, shoes, umbrellas, feathers and lace."[38]

Lovely things punctuate the monthly issues across all their genres; essays on literary history, interior design, travel, archaeology, and professions for women, as well as reviews and poems, repeatedly linger over objects and their sensuality. Illustrations were crucial for depicting the world of exquisite objects, and the lavish, high-quality visual texts in *Woman's World* make its signature style sharply different from competing titles, thereby suggesting that the commodification of things is attached to their visual appeal. As Wilde writes to the poet Alice Meynell in September 1887, requesting a contribution, "I prefer subjects that admit of illustrations."[39] And the visual appeal of feminized objects is also reiterated through the advertisements. In each issue, the contents page is cropped to half size, a layout that leaves room for more advertisements; in the February 1889 issue, the contents page shares room with an ad for other Cassell's publications. The advertisement papers also include pitches for miscellaneous feminine items, such as draperies, fabric, patterns, and gripe water. In May 1889, the eight preliminary pages of ads offer a dizzying variety of middle-class women's domestic and personal items: hats, a lemon squeezer, embroidery threads, undergarments, medicine, toothpaste, furniture cream, starch, books and weeklies, hosiery, shoes and plate powder, dresses, and Kodak cameras. After eight pages of advertisements, the half-page contents for the magazine is inserted, with room below for yet more ads. This is a layout (used each month) that neatly illustrates the literary contents' intimacy with commodities, for many of the contributions are concerned with precisely the objects advertised in the front and back papers: "Boots and Shoes," "A Lady of Fashion in 1750," "Quaint and Curious Needlework," and "Modern Gloves."[40] *Woman's World* is also a world of *things*, and expressly commodified things.

The magazine is quite literally wrapped up in advertising; for example, the end papers for May 1889 conclude with another flurry of advertisements for various feminine commodities juxtaposed with announcements of other books from Cassell's. The advertisements here, in fact, as with other months, are presented like miniature magazines within a magazine, with two pamphlets bound into the end papers (one for Pear's Soap and the other for women's personal items offered for sale by Philip Mason). The implication is that advertisements and the literary contents of the journal are both fetishized objects, and indeed each replicates the other. As scholars have noted, the magazine's investment in discussing objects as gorgeous commodities is part of its aestheticist agenda; it celebrates objects that connote feminized gender and under Wilde's editorship fashions its periodical personality as a commodity in itself.[41] And a crucial aspect of the magazine's poetry of commodified things is its regularly occurring poems.

Gillington's cycle of poems for 1889 appeared during a transitional time for *Woman's World*, and it bears a complex relationship to the editorial agenda. Wilde formally assumed his editorial role in November 1887 and, despite his boast to W. E. Henley that he worked only two days a week in the office at La Belle Sauvage for an hour a day and for six pounds per week, there is evidence that he took this role seriously—especially his obligation to attract society writers in order to boost the magazine's circulation.[42] He even invited the Queen to contribute. But the publishers were dissatisfied with flagging sales, and in October 1888 Wilde accepted a report from Cassell's chief editors, agreeing that the financial reality of the crowded women's magazine market meant that there had to be "more prominence . . . given to distinctly feminine subjects" (and, implicitly, that the protofeminist essays had to be toned down).[43] Wilde remained as editor until October 1889, the last time his name was on a monthly issue. The shilling magazine, as Wilde had stated in his response to the editorial review, was simply too expensive. Wilde's departure was a major blow for the magazine's aestheticist affinities, and *Woman's World* limped on without an editor named on the cover until its final issue in October 1890.

Gillington's cycle of pastoral poems is published during this period of falling magazine circulation, attempts to reassert the title's feminine personality, and the prelude to Wilde's departure. The poems' meditations on seasonal time and love's permanence, their highly mannered forms, and their visual impact as beautiful objects in their own right engage with the aesthetic Wilde forged in the magazine. The poems depict calendar time, rural time, and magazine time as the same, implying that the reader's world and the magazine's representation of a woman's world are identical—although such an equivalence is exposed as artificial in the interplay of the poem's stylized layout with the inky, rustic, picturesque illustrations that accompany it. The intricacy of the rhyming patterns in the poems—such as the villanelle, rondel, and triolet, old European literary forms fashionable in late-nineteenth-century aestheticism—emphasizes that the poems (like the intricate rhymes) mark time.

In the February valentine poem, for example, the three-dimensional realist sketch of the posy is placed over an antiqued paper

198 THE WOMAN'S WORLD.

consummation devoutly to be wished, yet the methods adopted by the so-called Socialists appear to me very far from adapted to the end they profess to have in view. True socialism, as opposed to individualism, appears to me to be more attainable by cultivating in ourselves and in others a spirit of unselfishness, order, and industry, than by artificially fostering an unreasonable hatred for their employers among the workmen of this country; a feeling which is so foreign to the naturally law-abiding temperament of the best English workmen, that I doubt if it will ever be put into them by the most eloquent and earnest of orators. It is a strange thing to go about preaching the gospel of brotherly love, and showing that the way to attain to it is to hate every one who is richer, or stronger, or wiser than you are yourself. Hatred, even of a common enemy, is but a poor cement wherewith to bind men together; the union it brings is but a name, as we have seen more than once in the various French revolutions. They began with vague worship of Liberty, which every one interpreted differently; of Equality, which only meant one dead level of guilt till one arch-hypocrite should have surpassed the others in crime; and of Fraternity, which was conveniently set aside to be taken up with bloody hands when the nation should be sick of hate.

Are not our own workmen far nearer the mark in their efforts to reach equality by raising themselves, instead of trying to drag down the rest of the world? They have learnt by bitter experience that, single-handed, they can hope to do nothing; they have learnt this lesson with a never-to-be-forgotten certainty that no amount of talk could ever have given them, and they have learnt it, too, without losing one jot of their healthy independence. So they have reached the conception and execution of a system of combination without which they were powerless. This has taught them many things, one of the most important being that combination of incapable and immoral men will do nothing; out of an infinite number of nothings you cannot make something, be the fusion and unity of these nothings ever so complete. What preaching would ever have taught that, do you suppose? You cannot help others if you cannot help yourself; before you can be of any use in an army you must be drilled, and in life there seems no unimpeachable authority but the one of circumstance.

Does all this sound like a plea for *laissez-faire*? It is not meant to be such. What I would say, rather, is that all progress and reform must come from within, not from without. Just as you will not strengthen a child's muscles by putting it in irons, so you will not strengthen a nation by forcing it to walk within the narrow walls of an iron law. It is the socialistic *spirit* which needs cultivating and spreading, not any one system of production or distribution to the exclusion of all others.

And now I hope it is clear that, if this be so, the sooner every man, woman, and child realises the true basis of society the better. Instead of Political Economy being the enemy of all charity and individual effort, it will be seen that no nation can be permanently, or universally wealthy, without love and co-operation.

HELENA SWANWICK.

St. Valentine's Day.

EVERY night I dream of
you,
Every morn at waking
Still the old deſire returns,
Still for you my whole heart
burns
With a love that knows
no ſlaking,
With a hunger ever new.

Birds fall ſinging where they
ſit;
Purple buds and roſy
Quicken on the hedgerow ſprays;
I have ſearched the woodland
ways
All to find a dainty poſy,
Something freſh and fair and fit.

Snowdrops white and celandine,
Firſt in field or garden;
Fearing, hoping, doubting ſtill,
Theſe upon your window-ſill
Here I lay—O grant me pardon,
Though you ſcout my Valentine!

Figure 7.1 M. C. Gillington, "St. Valentine's Day," *Woman's World*, February 1889, 198.

valentine, with a decidedly flat and unrealistic depiction of cupids, flowers, and flaming hearts; it is addressed "À ma belle / My heart," giving the addressee's name, "Amabel," its underlying meaning ("to my beauty/beloved"). The valentine asserts the textuality and artifice of the poem, but the realist posy also gestures to the referent. And the poetic form, with its mirrored *abccba* rhyme scheme, underlines the conventional and artificial mode that has a paradoxical claim to appearing like a notional, virtual world.

In the top right corner of the illustration is the banner for the month, adorned with curlicues reminiscent of the Renaissance examples of decorative art in *The Grammar of Ornament* (1856).[44] This banner prominently displays the poem's indexical link to the month (the pastoral month, as well as the magazine month) and adds to the antiquated ornamental style of the illustration's central frame, which is in tension with the naturalistic sketch of the posy of snowdrops and celandine. The striking simultaneity of antiquity and immediacy is echoed in the contrasts among the handwriting on the central illustration, the inscription on the valentine itself, and the typeface that deploys both a Gothic script and italics with a long *s* (other poems deploy *y*^e^ for "the"). This periodical poem mixes up its decorative visual and textual signifiers to encode both historicized pastoral and an episodic romance plot that is temporally aligned with the reader's consumption of *Woman's World.* As a decorative object in itself, the illustrated poem functions like other commodified things in the monthly magazine, simulating serial time as historical but also contemporary. Gillington's work has a cultural value based on its position within the serialized print media, where her cycle of poems signifies and punctuates the artificiality of serial time rather than serving as mere filler. Victorian periodical poetry refashions Victorian time, situating poetic writing and reading as interruptions in serial form, and also casting the pause as the source of serial print's intrinsic value. If, according to Mussell, Victorian periodicals only exist virtually, between each single issue, then poetry underscores the pause *between* installments and issues by figuring another space of pause *within* the issue. And, because Wilde is the editor, even this pause, as a beautiful thing, is part of the magazine's appeal to the consumer.

The pastoral mode of this cycle of poems intertwines stylization with the rustic picturesque. In each poem, the title is in Gothic type, the poem itself is in italics, and there is an adjacent picturesque illustration with the addition of a contrasting highly decorative feature (most notably the banner for the month). Each poem takes up the bottom half of the page, and each illustration is closely juxtaposed with its poem. (In some months, such as "The First Night of the Year," "St. Valentine's Day," "Gusty Weather," "Summer Night," "'All Among the Barley,'" and "A Complaint," the poem is even embedded spatially within the borders of the illustration.) The poems are laid out as set pieces, displays of artistry and value, beautiful objects in and of themselves, cultural treasures highlighted by the enclosing lines of the visual frame in most of the poems. The technology of the illustration, too, emphasizes verisimilitude and artistry. The photolithograph half tones, produced for *Woman's World* by a company in Vienna (Carl Angerer and Alexander Göschl), offer an image of photographic quality with many tiny dots, which from a distance gives the illusion of a continuous tone. In effect, the mechanical process of illustration simulates photographic realism, with the artifice of a decorative image. The representational strategies and technologies of Gillington's series of illustrated poems stand, as a synecdoche, for the larger world of the magazine; both create stylized, mannered, and supremely decorative realms that are also date-stamped with the temporal moment of reading and touched with the illusion of an indexical link to the real.

A triolet, "Gusty Weather," beautifully exemplifies the tensions within the visual/verbal field. The first letter of this poem for March, *W*, is part of the illustration, spelled in sticks in the field next to the figure of Amabel, a decorative letter that links the start of the poem to the illustration and that also starkly contrasts with the antique type of the title. This poem is an intricate and challenging triolet (with the rhyme scheme *ABaAabAB*, the capitals denoting the refrain), an eight-line form originating in medieval French and revived at the end of the nineteenth century after Gosse's 1877 *Cornhill* essay, "A Plea for Certain Exotic Forms of Verse," urged a return to the fixed forms of the triolet, rondel, rondeau, villanelle, ballade, and chant royal. These were to act as antidotes to the emotional and formal excesses of the Spasmodics, counteract "spontaneous and untutored expression," and elevate poetry "as one of the fine arts."[45] Gosse praises the triolet as "a very dainty little poem": "nothing can be more ingenuously mischievous, more playfully sly, than this tiny trill of epigrammatic melody, turning so simply on its own innocent axis."[46] Gillington's "Gusty Weather" reflects Gosse's playful tension between simplicity and artistry, naturalism and the antique, immediacy and an idealized rural past.

GUSTY WEATHER. 271

The Professor said not a word; but, before Mrs. Vernon could evade him, he caught her in his arms and kissed her, as ten years before he had kissed her on their wedding-day.

"Margery! My wife!" he cried at length as he held her from him to look into her eyes.

What he read therein must have satisfied him (and he was hard to please, as all his pupils knew), for he drew her into his arms again, and, being stronger than she, she could not get away.

"But you hate Chopin still," she said, at last; "and my rival, the microscope, is more formidable than ever."

"You may play Chopin from morning until night, if you will forgive me and love me again, Margery."

"You have more to forgive than I," the true woman answered, as she put up her hand and gently stroked the face which looked older and more marred than when she had seen it last; "I am ashamed when I think how horrid I was. But, let me whisper it, I thought you loved those dreadful worms and things better than your wife, and I was afraid I was too young, and——"

"And so we lost ten years of happiness," he answered, as her voice broke with a sob; but, although his voice was sad and rather stern, he stooped and fondly kissed away the mist of tears from her eyes.

It was some days before the Teesdales heard anything of the Professor. Then came a note from him, from the Isle of Wight. "I found Margery playing Chopin on the third floor," he wrote. "The microscope has a holiday, and we are enjoying a second honeymoon. I like it."

Quoth Mr. Teesdale, "She is a million times too good for that cantankerous little bookworm!" Quoth Mrs. T., "I am sure I hope she will be sensible now and treat him properly. Too good for him! Good gracious! Why, he is a perfect angel! I do not know a woman fit to tie his shoes."

Margery was of the same opinion, but she always tied the Professor's.

ELLA J. CURTIS.

Figure 7.2 M. C. Gillington, "Gusty Weather," *Woman's World*, March 1889, 271.

When Amabel a-milking goes,
  All in a kerchief grey,
If lighteſt wind but gently blows,
  When Amabel a-milking goes,

The dainty drifted folds difclofe
  A bosom white as May,
When Amabel a-milking goes,
  All in a kerchief grey.

Stedman, in *Victorian Poets*, calls triolets "little marvels" and defines Victorian poetry at the end of the century as the revival of lyric forms.[47] Gosse's plea for complex formal experiments to turn poetry back into art influenced many late-century writers, such as Austin Dobson and W. E. Henley; Richard Cronin dubs this turn in late Victorian poetry a "cultivation of difficulty" that "was a self-consciously anti-democratic stratagem."[48] And yet periodicals provided a crucial outlet for these formally refined poems, as well as for the reviews and essays that called for poetic artistry to be revitalized. Every poem in Gillington's cycle employs intricate stanzaic forms. Literary artifice is in tension with the naturalism of the illustrations and with the realist effect of the serial that equates magazine time with poetic time. Moreover, Gillington's *Woman's World* poems confirm that the poetic art praised by Gosse and his followers could be forged in popular, ephemeral magazines. Indeed, the cycle's use of strict French poetic forms as the vehicle for a standard novelistic romance plot, from courtship to marital pledge, suggests that disposable print was a promising medium—and middle-class female readers were a target audience—for formally sophisticated poetry.

Furthermore, particularly in the closed forms, Gillington's Amabel cycle uses rhyme to highlight poetry's investment in time as periodic, circular, and immediate. The triolets ("Gusty Weather" and "Amabel at Work"), the villanelle ("A Morning Meeting"), and the rondel ("A Complaint") each employ only two rhymes and are organized around repetition and refrain. As Adela Pinch notes of the fin-de-siècle turn to short rhyming lines, rhyme marks "passing time and making time."[49] Pinch, who focuses on the spare lyrics of Mary E. Coleridge, associates rhyme's twinned stopping of time and resistance to endings with late-nineteenth-century philosophical inquiries into time. In ephemeral and serial print media, time is also of the essence. And the frequency of seasonal poetry in periodical print throughout the Victorian era signals that such poems symbolically reflect much of the periodical's timeliness, while also offering (as Caley Ehnes remarks of the religious poetry in *Good Words*) a pause for reflection.[50]

Gillington's poem for August, "Amabel at Work," the second triolet in the series, deploys its two rhymes to celebrate and fetishize the body of Amabel, the speaker's beloved:

AMABEL AT WORK. 521

that according to his wish he was buried in the Church of Our Lady; his tomb remained unbroken down to the middle of the last century, and on his festival crowds of pilgrims came to see it, but not a vestige of it remains now. Glendalough remained a bishop's see until 1214, when it was incorporated in that of Dublin. Since then, this place, so sacred in its memories, has passed out of the region of history, and we hear no more of it.

Sir Walter Scott describes Glendalough as "the inexpressibly singular scene of Irish antiquities." There are other places in Ireland which witness to the same far-off time when monasteries and schools of learning abounded; but few unite in one, as this valley does, the wild scenery of mountain height and shadowy lake, with such dim but significant vestiges of grey and ruined buildings.

There are two very small but ancient churches close to the shore on the upper lake; they are so ruined that little is left but the walls and gables, but they are very striking in their green loveliness. There is one feature about these early churches which is very puzzling, and yet most characteristic of them, and that is their smallness. It is probable that the intense conservatism of the Celtic mind was one cause of this. From the introduction of Christianity the Celtic Church had been thoroughly monastic. We have seen that the monks, in place of occupying one particular building, established themselves in separate cells, with separate oratories, in which they performed their solitary devotions. The congregation, if there was one, was content to stand outside listening to the service as recited within, so it became the custom to build more than one little chapel in the same place. Ruined and deserted as these buildings are, they possess an intense interest to the antiquarian, as pointing to a time when Ireland, independent of foreign influence, was working out a style of architecture all her own, side by side with the development of that Christian faith which shone so brightly from her shores to the nations around.

It is easy, standing in the lovely valley of Glendalough, to recall the holy and peaceful life of that saint who made this place the home of prayer, the refuge of the scholar and the devout. Many a church and many a holy well have been dedicated to him in Ireland, but there is no place where his personality becomes so dear, and familiar to us, as in the valley where it is guarded by the mountain heights which he loved so well in life, and in the shadow of which he lies buried.

HONOR BROOKE.

Amabel at Work.

TRIOLET.

LAST evening when I paſſed yᵉ farm,
Sweet Amabel was making butter.
The cream ſplaſhed up her rounded arm
Laſt evening when I paſſed. The farm
In ſunſet light lay roſy warm;
Two white hands moved like doves a-flutter,
Laſt evening when I paſſed yᵉ farm,—
Sweet Amabel was making butter.

M. C. GILLINGTON.

138

Figure 7.3 M. C. Gillington, "Amabel at Work," *Woman's World*, August 1889, 521.

Last evening when I paſſed yᵉ farm,
Sweet Amabel was making butter.
The cream ſplaſhed up her rounded arm
Laſt evening when I paſſed. The farm
In ſunſet light lay roſy warm;

> Two white hands moved like doves a-flutter,
> Laſt evening when I paſſed ye farm,—
> Sweet Amabel was making butter.

The poem offers a conspicuous half rhyme ("farm / warm"), a break in the tight hyper-regularity of the triolet's conventional interlocking rhymes that is also signaled by the punctuation introduced into the repeated *a* rhyme in the fourth line. The effect is to emphasize "warm," suggesting the heat of the day's end, the speaker's passion for Amabel, and the labor of butter making. But our perspective is from inside the farm, not outside where the speaker presumably passes by. The sunset light in the illustration shines brightly in through the window from the direction of the poem's text, onto Amabel's face, arms, and upper body. The illustration catches her poised midaction with one hand in the cream, reminiscent of the everyday realism of a Vermeer painting but with the inky wash of late-century picturesque. The half rhyme signals this disjunction in point of view that both signifies realism and breaks with the speaker's perspective. Capturing the moment of voyeurism and desire, it also self-consciously arrests time for only an instant in the poem and, indeed, the series.

The August issue of *Woman's World* was the last month to publish Gillington's poems in a sequence that aligned poetic time with magazine time. This strange disruption in the seasonal immediacy of the cycle is hard to spot in the collated annual volume, in which the overt markers of a change in the month (such as divisions between parts, advertisements, and monthly contents) are removed. The last three installments in the series sever the poetry's deep connection with the monthly cycle of periodicity and seasonal time.[51] These contributions—which if following the pattern of one poem a month should appear in October, November, and December—are instead published out of synchronic order; what's more, the decorative banners for the month on their illustrations, which had been prominently filled in every other poem, are left empty. The poems lose their explicit link to the seasons and the months. The September issue publishes, along with the poem associated with that month ("All Among the Barley"), the poem presumably meant for October ("A Complaint"). The October issue publishes the poems presumably meant for November and December ("A Grey Day" and "Christmas Eve"). This loss of chronology and synchronicity is particularly striking given the previously consistent equation of magazine time with calendar time. For example, keen to promote the title's monthly hypertemporality, Wilde tells Violet Fane in January 1889 that her poem for the New Year, which he had just received, would look "positively *unpunctual*" if published at the next opportunity in March.[52] The February number had just gone to press as Wilde replied to Fane: evidence, if we needed it, that the magazine's hypertemporality was a careful construction. Wilde's last monthly issue with his name as editor was October 1889, and the collated annual volume for that year ran from November 1888 to October 1889. It seems probable that the dis-ordering of the rhythmical alignment of poems and months allowed Gillington to bring out all her poems under Wilde's editorship; after all, the future of the magazine was in question, and the poem cycle embodies closely the aesthetics of the magazine under Wilde.

While the banners for the month in "A Complaint" and "A Grey Sky" are empty, to signal their nonsynchronicity, the December medallion is presented differently. Once again the banner is empty, but—with images of ornate drapery, cupids, a heart pierced by arrows, and a scroll with wings tucked under a theatrical curtain—its artifice is contrasted even more sharply to a picturesque illustration of rural lovers. Perhaps December's empty banner signifies Wilde's departure from the virtual stage of *Woman's World*. In any case, the three empty banners disturb the balance in the layout and design of the verbal-visual text that hovers between the elaborately decorative and the picturesque. The blank labels reveal the constructed nature of the virtual periodical world.[53] And, underlining its nonsynchronicity while still attempting to resemble a calendar ending, Gillington's final poem, "Christmas Eve," is the last contribution to the last magazine issue of *Woman's World* edited by Wilde, and to his collated annual volume of 1889. *Woman's World*, the virtual world of women and their beautiful things and ideas, could not hold at a shilling a month in the late-nineteenth-century print market. It is apt that Gillington diversified her career into a variety of genres that also depended on time and timing but that had a more robust market and a more stable readership.

Periodical print culture marks time as supremely temporal, date-stamped, yet part of a continuing cycle. Periodical time is a stop and start or "progress and pause" (in Linda K. Hughes and Michael Lund's words).[54] As Mark Turner argues, nineteenth-century modernity was predicated on time as both newly regulated and inherently interrupted.[55] I want to extrapolate from Gillington's interrupted monthly cycle of poems to suggest that poetry in itself represents and visualizes the doubled time of the periodical. For not only is periodical poetry different textually and visually from the surrounding contributions, but it also represents *through its very poetic form* the pause as serial print's intrinsic value of immediacy and timeliness. Periodical poetry is an interruption in the serial form that is also deeply embedded within the serial's temporal cycles, suggesting that poetry can interact with this mode of print in ways that we have not yet begun to fully appreciate.

With the mass digitization of Victorian print, it might be tempting to make an analogy between the virtual world of periodical culture and the digital world that makes such print newly and easily accessible. But rather than situating Victorian print technologies as precursors to our own digital revolution, we need to examine the process of virtualizing print, the way in which Victorian poetry engages with an immersive world, and the methods by which poetry both disciplines and disrupts the hypertemporality of serial print. Victorian poets often expressed hostility to the ephemerality and heterogeneity of periodical print, and critics have tended until recently to view periodical poetry as filler that was secondary to serial prose fiction.[56] But poetry played a crucial role in the periodical, measuring and interrupting the rhythms of calendar time, exposing the mediations and constructions of the virtual world of print while also demonstrating its all-encompassing nature and simulation of the real.

NOTES

1. E. C. Stedman, *A Victorian Anthology 1837–1895: Selections Illustrating the Editor's Critical View of British Poetry in the Reign of Victoria* (Boston: Houghton, Mifflin, and Co., 1895) xi–xii; E. C. Stedman, "Victorian Poets," *The Century Magazine*, January 1873, 357–64.
2. E. C. Stedman, *Victorian Poets* (London: Chatto and Windus, 1875, revised and expanded edition 1887, 1893).
3. Oscar Wilde, "Literary and Other Notes," *Woman's World*, November 1887, 36–40 (39).
4. M. C. Gillington's 1889 poems in *Woman's World* are "The First Night of the Year" (January, 155), "St. Valentine's Day" (February, 198), "Gusty Weather" (March, 271), "Cloudy Skies" (April, 310), "The Shearing at the Stepping-Stones" (May, 381), "A Morning Meeting" (June, 405), "Summer Night" (July, 464), "Amabel at Work" (August, 521), "'All Among the Barley'" (September, 573), "A Complaint" (September, 610), "A Grey Day" (October, 632), and "Christmas Eve" (October, 664).
5. For an analysis of visuality, see Tim Fulford, "Virtual Topography: Poets, Painters, Publishers and the Reproduction of the Landscape in the Early Nineteenth Century," *Romanticism and Victorianism on the Net* 57–58 (February–May 2010); and Tom Gunning, "'We Are Here and Not Here': Late Nineteenth-Century Stage Magic and the Roots of Cinema in the Appearance (and Disappearance) of the Virtual Image," in *A Companion to Early Cinema*, eds. André Gaudreault, Nicolas Dulac, and Santiago Hidalgo (Oxford: John Wiley, 2012), 52–64.

6. Alison Byerly, *Are We There Yet? Virtual Travel and Victorian Realism* (Ann Arbor: The University of Michigan Press, 2013), 3.
7. Ibid., 6–7.
8. Hilary Fraser, Stephanie Green, and Judith Johnston, *Gender and the Victorian Periodical* (Cambridge: Cambridge UP, 2003), 200.
9. James Mussell, *The Nineteenth-Century Press in the Digital Age* (Houndmills, Basingstoke: Palgrave, 2012), 52.
10. Ibid., 56.
11. Ibid., 115.
12. See, for example, essays on the contemporary Woman Question, such as Julia Wedgewood's "Woman and Democracy," *Woman's World*, June 1888, 337–40, and Mary A. Marshall's "Medicine as a Profession for Woman," *Woman's World*, January 1888, 105–110. For an example of a historical essay, see Graham R. Tomson, "Beauty from the Historical Point of View," *Woman's World*, July and August 1889, 454–59, 536–41.
13. Here I want to acknowledge that my argument differs from Laurel Brake's interpretation of the magazine, which she sees as covertly homosexual. See *Subjugated Knowledges: Journalism, Gender and Literature in the Nineteenth Century* (Houndmills, Basingstoke: Macmillan, 1994), ch. 7.
14. M. C. Gillington, "Atlantis" and "The Home Coming," and Alice Gillington, "A West-Country Love-Song," in *Women's Voices: An Anthology of the Most Characteristic English, Scotch, and Irish Women*, ed. Elizabeth A. Sharp (London: Walter Scott, 1887), 405–08; Elizabeth A. Sharp, ed., *Sea-Music: An Anthology of Poems and Passages Descriptive of the Sea* (London: Walter Scott, 1887).
15. Mary C. Gillington, "A Dead March" and "The Home Coming," and Alice E. Gillington, "A West-Country Love-Song" and "Nocturnes," in *Women Poets of the Victorian Era*, ed. Elizabeth A. Sharp (London: Walter Scott, 1890), 278–86.
16. Edmund H. Garrett and Edmund Gosse, eds., *Victorian Songs: Lyrics of the Affections and Nature* (Boston: Little, Brown, 1895); Hugh Walker, *The Greater Victorian Poets* (London: Macmillan, 1895). For an analysis of Stedman's importance in defining national British and American literature, see Michael Cohen, "E. C. Stedman and the Invention of Victorian Poetry," *Victorian Poetry* 43, no. 2 (Summer 2005): 165–88. Under her married name, Mary C. G. Byron, Gillington's poems in *Victorian Anthology* are "The Tryst of the Night" and "The Fairy Thrall" (607–08); Alice's poems are "The Seven Whistlers," "The Rosy Musk-Mallow," and "The Doom-Bar" (608–09).
17. Leah Price, *The Anthology and the Rise of the Novel* (Cambridge: Cambridge UP, 2000), 9.
18. Oscar Wilde, "English Poetesses," *Queen*, December 8, 1888, *The Poetess Archive*.
19. David Doughan, "Queen (1861–1967)," in *Dictionary of Nineteenth-Century Journalism*, eds. Laurel Brake and Marysa Demoor (Ghent:

Academia Press and the British Library, 2009), 523–24 (524). See also Margaret Beetham, *A Magazine of Her Own? Domesticity and Desire in the Woman's Magazine, 1800–1914* (London: Routledge, 1996), ch. 8.

20. Kathryn Ledbetter, "Time and the Poetess: Violet Fane and the Fin-de-Siècle Poetry in Periodicals," *Victorian Poetry* 52, no. 1 (Spring 2014): 151.
21. Merlin Holland and Rupert Hart-David, eds., *The Complete Letters of Oscar Wilde* (New York: Henry Holt and Company, 2000), 210.
22. Richard Ellman, *The Consciousness of Joyce* (London: Faber and Faber, 1977), 59–61; Molly Hoff, "The Pseudo-Homeric World of *Mrs. Dalloway*," *Twentieth-Century Literature* 45, no. 2 (Summer 1999), 186–209 (192, 205). Thanks are owed to my colleague J. Matthew Huculak, who offered these references.
23. "M. C. Gillington," *The Musical Herald*, August 1, 1904, 227–29 (228).
24. See, for example, *May Byron's How-to-Save Cookery: A War-time Cookery Book* (London: Hodder and Stoughton, 1915), *May Byron's Rations Book* (London: Hodder and Stoughton, 1918), and her *Pot-Luck, or the British Home Cookery Book* (London: Hodder and Stoughton, 1914), which reached at least seven editions.
25. Joseph Bristow, ed., *The Fin-de-siècle Poem: English Literary Culture and the 1890s* (Athens: Ohio UP, 1995), xxvi.
26. Eveline Portsmouth, "The Position of Women," *Woman's World*, November 1887, 7–10; Violet Fane, "Hazely Heath," *Woman's World*, November 1887, 10. See Ledbetter, "Time and the Poetess" 151. I am indebted to Ledbetter's work, which has shaped my conception of late-century periodical poetry.
27. Amy Levy, "The Recent Telepathic Occurrence at the British Museum," *Woman's World*, November 1887, 31–32; unsigned, "The Oxford Ladies' Colleges," *Woman's World*, November 1887, 32–35; Oscar Wilde, "Literary and Other Notes," *Woman's World*, November 1887, 36–40.
28. Wilde, "Literary and Other Notes" 38.
29. Ibid., 39.
30. See the "Literary and Other Notes" for December 1887, 81–82; February 1888, 182–83; December 1888, 111–12; March 1889, 277–78; May 1889, 389–90; June 1889, 447–48.
31. Wilde, *Letters* 332.
32. The monthly parts of *Woman's World* were 48 pages long, in contrast to the 36 pages of *Lady's World* (Brake, *Subjugated Knowledges* 139). In November 1888, the magazine ran a full-page advertisement to announce that, at no extra cost, it would increase in length.
33. Wilde, *Letters* 297.
34. "Christmas Books," *The Times*, December 7, 1888, 13.
35. "The Woman's World," *The Spectator*, December 7, 1889, 9.
36. This quotation is from Wilde's "The Soul of Man Under Socialism" (1891), in which he imagines a future free from private property and the accumulation of things. See *The Complete Works of Oscar Wilde*, vol. 4, ed.

Josephine Guy (Oxford: Oxford UP, 2007), 238–39. Green situates this comment from Wilde in relation to his earlier editing of *Woman's World*, as an example of his general equivocation about object possession (116–27).

37. Wilde, *Letters* 301.
38. Stephanie Green, "Oscar Wilde's *The Woman's World*," *Victorian Periodicals Review* 3, no. 2 (Summer 1997): 102–20 (116, 117).
39. Wilde, *Letters* 320.
40. B. De Montmorency Morrell, "Boots and Shoes," *Woman's World*, May 1889, 343–48; Alice Comyns Carr, "A Lady of Fashion in 1750," *Woman's World*, May 1889, 373–77; Ellen T. Masters, "Quaint and Curious Needlework," *Woman's World*, May 1889, 382–85; S. William Black, "Modern Gloves," *Woman's World*, May 1889, 386–88.
41. As Green notes of women's magazines generally (116).
42. See Anya Clayworth, "*The Woman's World*: Oscar Wilde as Editor," *Victorian Periodicals Review* 30, no. 2 (Summer 1997), 91.
43. Wilde, *Letters* 363.
44. Owen Jones, *The Grammar of Ornament* (London: Bernard Quaritch, 1868 [1856]), ch. 17.
45. Edmund Gosse, "A Plea for Certain Exotic Forms of Verse," *The Cornhill Magazine*, July 1877, 53.
46. Ibid., 62.
47. Stedman, *Victorian Poets* 474.
48. Richard Cronin, *Reading Victorian Poetry* (Oxford: John Wiley, 2012), 205.
49. Adela Pinch, "Rhyme's End," *Victorian Studies* 53, no. 3 (Spring 2011): 485–94 (486).
50. Caley Ehnes, "Religion, Readership, and the Periodical Press: The Place of Poetry in *Good Words*," *Victorian Periodicals Review* 45, no. 4 (Winter 2011): 466–87.
51. Other magazines of the nineteenth century, such as *The Cornhill Magazine*, had also attempted to forge a connection between their metropolitan bias and rural life.
52. Wilde, *Letters* 386.
53. Cassell's was based in this period at La Belle Sauvage, and it is possible that the play in Gillington's poems with pictorialism and artifice is a reference to the magazine's offices. For an account of the name, see http://www.londononline.co.uk/streetorigins/La_Belle_Sauvage/.
54. Linda K. Hughes and Michael Lund, *The Victorian Serial* (Charlottesville: UP of Virginia, 1991), 63.
55. Mark Turner, "Time, Periodicals, and Literary Studies," *Victorian Periodicals Review* 39, no. 4 (2006): 309–16.
56. Linda K. Hughes's call for periodical poetry to be taken seriously is a watershed in criticism: "What the *Wellesley Index* Left Out: Why Poetry Matters to Periodical Studies," *Victorian Periodicals Review* 40, no. 2 (2007): 91–125. See also Alison Chapman and Caley Ehnes, eds., *Victorian Periodical Poetry*, special issue of *Victorian Poetry* 52, no. 1 (Spring 2014).

CHAPTER 8

ARTIFICIAL ENVIRONMENTS, VIRTUAL REALITIES, AND THE CULTIVATION OF PROPENSITY IN THE LONDON COLOSSEUM

Peter Otto

> *"Presence" . . . is the first word that should come to mind in understanding the changes that are now taking place as art becomes science. What if presence can be re-engineered? Certainly, that is what is being attempted.*
>
> —*Nigel Thrift, "Pass It On"*[1]

The panorama is routinely identified with the modern attempt to contain everything within a single view or picture—and as such, with one of the earliest steps in a narrative that leads, by way of the diorama and the invention of photography, to the moving images of the cinema. This narrative is often thought to culminate in the digital virtual realities of our own era, which add "navigation, immersion, and interaction to the cinematic representation";[2] yet this same contemporary development disrupts the narrative it appears to conclude. By recalling the panorama's immersive virtual realities, digital virtual realities foreground a second narrative, an unruly counterpoint to the first, which is driven by a logic of immersion and participation rather than detachment and representation, is more closely linked to romantic than to Enlightenment thought, and suggests a significantly different relation between nineteenth- and twenty-first-century virtual realities.

In the following pages, I want to sketch some of the contours of this second narrative, through a discussion first of the allocentric and autocentric dimensions of illusion in early panoramas, and then of Thomas Hornor's Colosseum, the "Panorama of London," which it housed, and the virtual realities that it constellated. If measured by the degree of verisimilitude they achieve, the Colosseum's virtual realities are rivaled only by modern theme parks and computer-mediated immersive experiences. And yet, as I will argue, this hyperrealism is, in the early panorama, not finally divisible from the use of space to produce mood or atmosphere, which will in turn cultivate propensity, understood as "a disposition to behave in" ways that are "only partly in the control of the agent."[3] This shifts attention from the real world back to the romantic artist as a creator of worlds and to the panorama's audiences as both shaped by and cocreating these worlds, the elements of which form the material for a rich "second life."

Natural Object and Panoramic Affect

The most common response to late-eighteenth- and early-nineteenth-century panoramas was astonishment, prompted by their power to present historical events or distant locations as if they were actually before the eyes of their audiences. A reviewer for *The World*, for example, after visiting Robert Barker's "View of Edinburgh" (1789), the world's first full-circle panorama, predicted that

> more trips to Scotland will soon take place, than has been done at any preceding season . . . an ingenious Artist having contrived to bring not only the Capital of that Kingdom, but also an extensive circle of the surrounding country, into the Haymarket. There seems nothing now wanting to complete the felicity of the Masters and Misses, but the noted Blacksmith from Gretna-Green.[4]

With this degree of realism, virtual tourism was a real possibility. As a reviewer for *The Times* remarked two weeks later:

> Mr Barker's improvement in Painting . . . must prove particularly interesting to their Majesties . . . who rarely go abroad. To them views of distant countries will be brought . . . before them, one entire uninterrupted circle, placing them in the centre, where they can see the same as those who travel; . . . and having seen it personally, they can retain it perfectly in idea, the same as nature could impress.[5]

The "View of Edinburgh" was painted on a cylinder only 25 feet in diameter and displayed in a room not designed for the purpose. With the opening of Barker's panorama rotunda, which was 90 feet in diameter, in Leicester Square on May 25, 1793, the panorama's illusions became still more lifelike, and virtual tourism became an everyday reality. This new industry began at home, with Londoners fascinated by Barker's "view-at-a-glance of the CITIES OF LONDON and WESTMINSTER" as seen from the top of the Albion Mills, which was exhibited from June 1791 in Castle Street near Leicester Square.[6] More exotic views soon followed. As Ephraim Hardcastle wrote in the *Somerset House Gazette* (1824):

> We have seen Vesuvius in full roar and torrent, within a hundred yards of a hackney-coach stand . . . Constantinople, with its bearded and turbanned multitudes, quietly pitched beside a Christian thoroughfare . . . and now Pompeii, reposing in its slumber of two thousand years, in the very buzz of the Strand.[7]

Although Hardcastle considers George Berkeley (1685–1753) a mere "metaphysician," whose "word" counts "for nothing but waste of brains, time, and printing-ink," the panoramic illusion seemed to prove the most outrageous of this philosopher's claims, the foundation of his idealism—namely, that *esse est percipi* ("to be is to be perceived"):

> There is no exaggeration in talking of those things as really existing . . . If we have not . . . the bricks and mortar of the little Greek town, tangible by our hands, we have them tangible by the eye . . . The scene is absolutely alive, vivid, and true; we feel all but the breeze, and hear all but the dashing of the wave.[8]

With this degree of agreement between appearance and reality, it is hardly surprising that the panorama was, and continues to be, more closely aligned with allocentric than autocentric modes of art, and therefore associated with Enlightenment rather than romantic thought. Like the "true philosophy" imagined by Francis Bacon (1561–1626), it seemed to have been "written as it were at the world's own dictation; being nothing else than the image and reflexion thereof, to which it adds nothing of its own, but only iterates and gives it back."[9]

Who would deny the realism of the panorama? And yet, as is no doubt already evident, the narrative that has brought us to this point is closely entwined with a second, which brings back inside the panorama phenomenon much of what the other excludes. Like the first, this second narrative begins with the hyperrealistic illusion conjured by Barker's "View of Edinburgh"—but rather than remaining spellbound by the congruence between panoramic illusion and reality, it shifts attention to the virtual reality itself, the artists that frame and the machinery that supports its existence, and the influence it exerts on the audiences who bring its world to life each time they enter its spaces.

In the account of Barker's panorama published in *The World*, for example, interest in the panorama's realism soon shifts to fascination with its ability to collapse time and space, the opportunity to travel without traveling that this affords, and the objects of desire it brings within reach. Indeed, when the reviewer mentions Gretna Green, a Scottish village just across the border from England, realism is no longer in his sights. After the passage of Lord Hardwicke's Marriage Act in 1753, which made formal marriage ceremonies mandatory and which applied to England but not Scotland, Gretna Green became famous as a place where lovers could still marry clandestinely, minors could marry without the consent of their parents, and irregular marriages (popularly thought to be performed by the village blacksmith) were still recognized.[10] When the reviewer aligns the panorama with Gretna Green, and the former's audiences with the latter's "Masters and Misses," he is therefore playfully identifying both as spaces of ungoverned desire; lying just beyond the reach of the status quo, such spaces allow romantic fictions, licensed by the artist/blacksmith and in which spectators/lovers play the leading parts, to become tangible realities.

The reviewer for *The Times* veers just as decisively from the panorama's verisimilitude, by focusing on the way it alters the scopic regimes of late-eighteenth-century England. Panoptic power is conventionally reserved for kings, who announce its presence through the display of maps and cartographic globes. But in the panorama this gaze is offered to the king by the artist, rather than by God, and is shared by anyone willing to pay "half-a-crown" for the privilege.[11]

The effects of this space of desire, which eludes the grasp of authority and reality while making authority's gaze and reality's objects available to all, are the subject of Hardcastle's reflections.[12] Rather than remaining inside the panorama, the artist's hyperrealistic illusions have moved outside, altering the texture of reality. Displacing the world created by God, ruled by the King, and mapped by science, London has become a dream space in which the virtual and the actual, the past and the present, the distant and the near, the objects of subjective desire and of objective reality, have moved into proximity with each other. On the one hand, the interstices of the real overflow with other times, spaces, realities, histories, and events, which seem to be "really existing." And on the other hand, as these virtual worlds proliferate, they tug at what had seemed to be substantial reality, raising the possibility that it might itself be a virtual reality and that the "natural *objects of our perception*" might be no more than "*virtual* appearances."[13] As Hardcastle suggests, one of the primary lessons of the panorama, which recalls the paradoxes of Berkeley's immaterialism, is that perception does not need its apparent object.[14] But what then of the panorama's realism?

I am, of course, not about to suggest that panoramic painting should be classified as a loosely idealist rather than a strictly realist genre, fathered by Berkeley rather than Bacon. To do so would obscure its most remarkable feature—namely, the extent to which, for those within its spaces, allocentric and autocentric domains overlap with each other, with each providing the ground on which the other depends. As a simulation, the panoramic illusion can be experienced as a re-presentation of *and* an alternative to the real, a space tied to *and* distanced from objective reality, which appeals to our reason *and* our desire. Although the panorama's viewing platform might on occasion gather disinterested observers, it was more commonly a scene of exchange between "social actors" (it could hold 150 at a time, according to Stephan Oettermann),[15] who were immersed in and engaged by painterly *and* social scenes. The pact between scientist and artist is therefore always already entangled with the alliance between artist and entrepreneur. When in league with scientists, panorama artists have verisimilitude as their goal and the detached observer as their audience, but when they join forces with the entrepreneur, they become more interested in the construction of immersive space and the hold it exerts on audiences. In nineteenth-century London, the alliance (and the tension) between these tendencies can most vividly be seen in the Colosseum, the panorama it housed, and the virtual realities with which they were associated.

CONJURING THE REAL

At Barker's Leicester Square panorama rotunda, only a few steps divided the outside world from the viewing platform and panoramic illusion; but in the Colosseum, audiences stepped first into the

Saloon—a circular room, 100 feet in diameter, which displayed "a collection of pictures, sculptured and fancy pieces"[16]—and only then into the "ascending car," contained by the hollow column at the center of the Saloon, which took them to the panorama itself. One might say, quoting William Hazlitt (1778–1830) out of context, that like all "fine [galleries] of pictures," the Saloon "is a sort of illustration of Berkeley's Theory of Matter and Spirit. It is like a palace of thought—another universe, built of air, of shadows, of colours."[17] But Hazlitt's other universe is entered only in thought or imagination, whereas when audiences stepped into the first of the Colosseum's four galleries, two things happened: first, it was as if they had themselves become figures inside the virtual space conjured by one of the Saloon's paintings; and second, the world seemed to turn itself inside out, bringing the whole into the part, the outside into the inside, and the infinite into the finite space of the Colosseum. In this remarkable moment people who had previously been spectators and/or mental travelers ("imagonauts") became immersants, who, in their first moments in this virtual world, were overwhelmed by a flood of sensory information—the "grand burst of magical and imposing effect" that "breaks upon [the] bewildered eye."[18]

After a few moments, when the image came into focus, immersants found themselves standing on what seemed to be the summit of St. Paul's, with the cathedral's dome curving away beneath them. They were at the center of a view that, unfolding beneath the vault of the sky (painted on the dome of the Colosseum), extended to the distant horizon and encompassed "the dwellings of nearly a million and a half of human beings—a countless succession of churches, bridges, halls, theatres, and mansions—a forest of floating masts, and the manifold pursuits, occupations, and powers of its ever-active, ever-changing inhabitants."[19] Reviewers reported again and again, "It seems scarcely possible for painting to achieve anything nearer to reality than has been effected."[20]

Those wanting to test this second reality could climb to the gallery built on "the summit of the building"[21] and compare the actual with the virtual prospect of London they had just seen. This took audiences back to the actual world, but with an important difference—namely, that the real was now seen as a view in competition with the unreal. As John Britton remarks, "the spectator will be gratified in comparing the colouring, perspective, and effects of nature, with those of art which he had previously examined: and he will then be disposed to award to the latter its due share of merit and applause."[22]

VIRTUAL WORLDS AND SYMPATHETIC IMITATION

Hornor's most obvious attempt to thematize the competition between nature and art, actual and virtual worlds, could be seen above the Colosseum's third gallery. Here he displayed, in ascending order, the copper ball that had been "placed upon [the summit] of St. Paul's by Sir Christopher Wren";[23] a "facsimile of the cross by which it was surmounted"; and—hanging above the dome, ball, and cross of St. Paul's—the wooden cabin in which he had made the drawings on which the "Panorama of London" was based.[24] This ensemble associates the panorama painter with Satan and his democratic artistic vision with the third of Satan's temptations of Jesus, which takes place on "a pinnacle of the temple" (Luke 4:9–12). As in romanticism more broadly, Satan's rebellion against the "Monarchy of God" and his creation of Pandemonium[25] here become figures for the artist's creation of a virtual world, which he places in competition with the actual world created by God. And although in the Colosseum it is the ascending room (rather than Satan) that carries everyday audiences (rather than the Son of God) to a view of London (rather than Jerusalem), the temptation derives from the same phenomena—namely, the pleasures offered by a world centered on the human rather than the divine and available to those willing momentarily to loosen their ties to the existing order of things. Indeed, one might say that as spectators stand on the summit of the dome of the Colosseum (above the summit of the virtual St. Paul's), this is precisely the choice they are playfully being asked to make.

If one's gaze is focused on the allocentric dimensions of the panorama, these last remarks will sound implausible. And yet, rather than resembling an intermediary that innocently transports the contours of the real world to its audience, the panorama is closer to what Bruno Latour describes as a *mediator*, with its own powers to "transform, translate, distort, and modify the meaning of the elements [it is] supposed to carry."[26] In contrast both to the actual world, structured by all manner of borders (inclusions and exclusions), and to the carefully framed views of conventional paintings, the panorama had no principle of "composition other than that *implied* by the chosen vantage-point," and "neither knew how to, nor did it want to select."[27] It therefore conjured "an environment or milieu . . . filled with a relatively unstructured profusion of visual data, unmanageable in its detail."[28]

This democracy of (virtual) objects is mirrored by the human democracy formed on the panorama's viewing platform. The views from the internal and external galleries of the Colosseum, like those offered by other panoramas, can be associated not only with the panoptic power normally reserved for kings but also, more broadly, with a masculine and aristocratic point of view, "which on grand estates is symbolized by the stately home."[29] But in the panorama, the stately home has become a viewing platform where this outlook is available to all. This collocation of spectatorial freedom and unstructured environment established "a field of open causality, each determination of which is an enactment by the spectator of only one of the possibilities it contains."[30] And this in turn generated a sense of relative freedom and arguably also of agency, in excess of that available in everyday life, which together help explain "why [the panorama's] popularity grew so swiftly and lasted for so long."[31]

Although the preceding paragraph does not introduce a postmodern picnic, in which the painter brings the objects and the spectator brings the frames, it is true that the roles open to viewers, and the ways in which the painting could be parsed, were remarkably diverse. Two of the most popular roles were "awed spectator" and "proud patriot." Framed in this way, Hornor's panorama evoked the wisdom of Britain's government, the modernity of its commercial culture, and the grandeur of its political institutions, along with a sense of plebeian political and artistic freedom. In the words of one of the panorama's first reviewers:

> The scene gives rise to so many inspiring associations in an enthusiastic mind, that few Englishmen, and still fewer Londoners, are equal to the detail of its description. Every inch of the vast circumference abounds with subject for reflection. The streets filled with passengers and vehicles—the grandeur of the public buildings, churches, and palatial structures—the majestic river winding grandly along . . . till it stretches away beyond the busy haunts of industry, to the rural beauties of Richmond, and the castellated splendour of Windsor.[32]

The attraction of the panorama here derives from the link *and* mismatch it engineers between the audience's actual experience and their virtual view of London, with the latter rearticulating the first in ways that place audiences within, and offer them a vicarious share of, the sublime power animating the nation as a whole.

The disparity between the actual and the virtual evoked by a particular panorama could, of course, be construed in dramatically different ways. A few weeks after the review we have been discussing was published, a critic for the *London Magazine* asked readers whether they could

> see that dark-looking building, and its narrow inner courts, a little to the right of the north-western pinnacle of the Cathedral? Did you think Newgate was such a straitened place? And yet three thousand prisoners have passed into its dreary walls, and the greater number have passed out to banishment, and a few to death, during the last year . . . Look to the North of Cheapside, where there is a huddle of miserable hovels. That is Spitalfields.[33]

Seen from this perspective, the "magic" of the Colosseum, which turns illusion into apparent reality, is matched by Westminster Abbey and the Hall of Rufus, both clearly visible in the painting: "Great are the mysteries transacted beneath that roof; and violent are the transformations of the palpable into the obscure, of truth into fiction, of fiction into truth." And this "magic" inflects what can be seen in the panorama: although St. Stephens Chapel was, from 1547 to 1834, the meeting place of the House of Commons, it can barely be discerned:

> crammed in between Westminster Hall and the House of Lords . . . Between the privileges of the Aristocracy and the precedents of the Judicature, it would be out of reason that St. Stephen's should make much figure in the panorama of London—so give over looking for it.[34]

As these divergent responses suggest (and they could in number and type be multiplied many times), the differential between the actual and the virtual opened by the panorama generated passion in its audiences, rather than simply offering an opportunity for passionate expression. We can understand why by turning to John Locke's argument, in *An Essay Concerning Human Understanding* (1690), that the "discrepancy between one's 'idea of delight' and the object whose possession would provide the sensation of delight" is sufficient to bring the entire spectrum of the passions into being, from joy and sorrow, to hope, fear, despair, anger, envy, and shame.[35] As this concept develops through the eighteenth and early nineteenth centuries, in the work of Edmund Burke (1729–97), Adam Smith (1723–90), and Dugald Stewart (1753–1828), it becomes evident that "ideas" gain their hold on us in large part through empathy or sympathetic imitation, which causes our bodies to be moved as if what is happening (or represented as happening) to others were happening to us.

Although accounts of the panorama often imagine the phenomenon as it might have appealed to the spectator's reason, its primary power comes from its ability to affect the bodies of immersants and, in so doing, to create a mismatch between "idea" (the virtual world) and object (the actual world within which audiences live their everyday life).

Inside the panorama, this primary imitation and the disparate emotions it provokes become the catalyst for secondary imitation—as immersants look around to see how their own responses are being viewed by others, as well as how those others are reacting to the panorama and to their companions, they are manufacturing the *prima materia* for social exchange. No doubt something similar occurs in ordinary social exchanges and at theaters and art galleries, but the hyperrealistic and hypersocial environment of the panorama intensifies both phenomena. Indeed, one could say that the opportunity for primary and secondary imitation is the principal pleasure retailed by the panorama, the chief source of its popularity, and, as I will suggest in the next section, the foundation of the "second life" it offers consumers.

Inside the Colosseum's Second Worlds

Most panorama paintings can be viewed only from a single platform, but at the Colosseum there were three internal galleries, which together created the sense that immersants could move around inside its virtual world. And contrary to the truism that panoramas lack frames,[36] the author of *A Picturesque Guide to the Regent's Park* explains that the front of the Colosseum's first gallery was divided by "pilasters and an entablature," and the second and third by "piers supporting arches." These formed "frames" that gave the galleries "an architectural character," while also tending "materially to assist the pictorial effect of the panorama," by enabling the visitor "to contemplate separately and uninterruptedly any particular portion of the extensive scene before him."[37]

The rationale for these important adjustments becomes evident when we turn to the third innovation, which develops the Colosseum's viewing platforms as sociable spaces. All panorama viewing platforms are sociable spaces, it would be reasonable to remark; but at the Colosseum, they are meant to draw people together for more extended periods of time. The space at the center of the second gallery, for example, was designed as a refectory,[38] while a room built on the same level as the ball and cross was intended for music and balls.[39] These spaces were supplemented by the Saloon, which was planned as a site for social interaction, and by a "beautiful reading-room, on the north side of the Colosseum, with French windows and rusticated Gothic verandas."[40] Hornor intended the "reading-room" to be part of a suite of rooms, which included a library and 30 smaller apartments, and which together would provide "to subscribers all the advantages of a club."[41] Saloon, refectory, ballroom, reading room, library, private apartments, and club—it is almost as if a new kind of stately home is taking shape, big enough to accommodate large portions of the public and with views more expansive even than those shaped by the landscape designer Humphry Repton (1752–1818). These experiments in the creation of sociable space were taken a step further by the fourth and arguably the most remarkable of Hornor's innovations, which was to multiply the number of viewing platforms, dream prospects, and virtual worlds available to immersants.

After audiences had been carried by the "ascending car" back to the Saloon, most moved on to the Colosseum's Swiss cottage, designed by the architect Peter Frederick Robinson (1776–1858), where the open, relatively unstructured social spaces of the panorama's viewing platforms were exchanged for "a series of small chambers." These offered a view of a wild rather than civilized environment, composed of "valley and hill, rock and cataract, pine forest, glaciers, and snow-capped Alps."[42]

The viewing platforms of the panorama and Swiss cottage opened to vast external realms; but when audiences moved to the conservatory, reached from the Swiss cottage through a subterranean passage, the outside seemed to have been brought *inside* the viewing platform. This structure, the first of its kind in Britain,[43] was divided into six different apartments, forming a promenade that extended for nearly three hundred feet.[44] Outlined by light wrought-iron frames, which supported panes of glass, it was filled with exotic "plants and flowers"—such as the *Camellia japonica* at the center of one of the apartments, which was thought to be "the most magnificent specimen in England."[45] But its most often-remarked feature was the fountain, placed at the center of an apartment shaped as "a large and lofty dome," which came into sight as audiences arrived from the Swiss cottage. It included "a circular basin," bordered with "shells and corals,"

> immediately within which is a continued row of jet-d'eaus. These throw a sort of transparent veil or mantle of water high in the air, with an inclination to converge and fall in the centre upon a columnar mass of shells, corals, and mosses. Near the top of this mass is a sort of dial of shells, which continually revolves by the action of invisible machinery, and which, combining with the spray from the falling waters, produce many beautiful prismatic effects when the rays of the sun glance on it in certain positions.[46]

The fountain's watery veil, "invisible machinery," and constantly changing optical effects provide an obvious allegory of the varying

THE PRESENT FASHIONS. B. READ. 12, Hart Street, Bloomsbury Square, LONDON.

Figure 8.1 *The Present Fashions*, published and sold by B. Read, 12 Hart Street, Bloomsbury Square, London.

spaces and appearances of the conservatory, the virtual reality housed beneath the much larger pleasure-dome of the Colosseum, and the ensemble of virtual realities constellated around it. But it also provides an allegory of the human display that unfolds in the Colosseum's sociable spaces.

In the print entitled *The Present Fashions*, for example, published by the fashionable tailor Benjamin Read in 1830, the fountain appears in the background of the design, as if it were the machine from which what we see has been projected.[47] This effect is heightened in three ways: first, by the visual cone, with the fountain at its apex, outlined by the pillars and arches stretching from the background to the foreground of the design; second, by the triangle, also with the fountain at its apex, which is suggested by the arrangement of large standing urns and human bodies; and third, by the elements that link the fountain with the urns and the women in front of it. The "dial of shells" that crowns the fountain, for example, is echoed by the huge bouquets of flowers (roses, daylilies, delphiniums, and coneflowers) that crown the large standing urns, placed on the right- and left-hand sides of the conservatory, and by the hats worn by the women (crowns this time of fabric and ribbons). Further, the body of the fountain echoes the bodies of the urns, while the fountain's torso (comprising a backbone and two arms formed of shells), watery veil, and rounded base are echoed respectively by the women's bodies, veils, and dresses.

The visual cone is contrasted with a second, seen in the background of the design. Its almost-vertical wrought-iron ribs rise to an apex immediately above the fountain, just beyond the upper border of the print, giving the image a second vanishing point. This establishes a contrast between the horizontal lines of the first cone, which point to the fountain inside the conservatory, and the upward-reaching lines of the second, which lead the viewer's eye to an invisible, immaterial point outside and above the design. The rounded form of this second cone recalls the domes of St. Paul's and of the Colosseum, while its vanishing point can be associated with the pinnacle of the temple—which is the domain of the masculine and aristocratic observer, who stands far above the worlds he is viewing, but also of Hornor/Satan (the artist-engineer as creator), who has constructed the virtual worlds of the conservatory and of the Colosseum as a whole. The contrast between cones therefore quickly becomes a contrast between male spectator/artist/engineer and female spectacle, with the former lifted above the scene in which women are immersed—although one must add that the design establishes these distinctions only to muddy them.

Inside the first cone, as we have seen, a sequence of visual echoes links the fountain, urns, and floral displays with female spectacle. In the second cone, an analogous sequence traces a path from outside and above the conservatory into its midst: the rising lines on the glass dome, which converge at its apex, parallel the cane-like stems of the tall plants (probably *Dracaena marginata*, from equatorial Africa and Asia), which hold heads of swordlike leaves aloft above the crowd and urns. And these rising wrought-iron ribs and trunks parallel the vertical form of the men's bodies, accentuated by their dark-colored clothes, which hold aloft their whiskered faces and elongated hats.

The divergence between male and female viewing positions thus far remains in place: rising ribs, cane-like trunks, top hats, and masculine bodies contrast strongly with rounded arches, capacious urns, wide-brimmed hats, and feminine bodies. And yet, at the same time, the men are also immersed in the virtual world of the Colosseum and have become part of its display: in this environment men and women stand on the same ground, where they look at each other *and* look at each other looking. The woman standing in the center-left foreground of the design, for example, sports a feminine version of the top hats worn by the men; like her partner, she holds a walking cane in her hand (although her eyes and hands draw our attention to an object of desire very different from the one that has caught the attention of her friend). At the far right of the design, a man looks back at us, as if we too have been drawn from the outside into the midst of this virtual scene and become part of its spectacle. Fountain, urns, flowers, fabrics, ribbons, trunks, crowns of swords, and male and female bodies and fashions together comprise a cornucopia of delights, of constantly changing sensuous surfaces, which the design presents as spilling out from the conservatory into the domain of the viewer.

But this is still only a partial summary of the design, which also traces the multilayered virtual world that rises from these surfaces. The foundation of this world is suggested by the children in the lower corners of the print, who express a simple pleasure in illusion: one points to the passage to the Swiss cottage, which will take him to another world; the other plays with a wooden creature that, like the Colosseum's other illusions, almost seems real. The degree of their absorption in virtuality is suggested by their gazes, which ignore or look past their human companions.

Although similar illusions texture the world in which the adults are standing, they are a disenchanted and self-reflexive audience, for whom the pleasure of visual trickery is intensified by the knowledge that they "are in raptures thanks to the perfection of an artifice."[48] This introduces the next level of the design, in which the audience's raptures become the object of sympathetic imitation and the subject of social exchange—the vigor of which is suggested by the number of different postures, gestures, glances, and expressions animating the crowd.

And finally, social display, sympathetic imitation, and social exchange bring bodies into new relation with each other. As noted above, according to the review of Barker's "View of Edinburgh" published in *The World*, all that was required "to complete the felicity of the Masters and Misses [was] the noted Blacksmith from Gretna-Green." The same might be said here regarding the young man and woman on the right-hand side of the design, immediately to the left of the large urn. Their flirtation is a synecdoche for this scene's sexual frisson, evident in looks, gestures, and the play between circular and rising forms that holds the assembly together. This frisson is unlikely to be any less when the crowd moves from the conservatory into the surrounding gardens, where Hornor had exercised the "necromantic, or talismanic power" of his art to create "mountains, dells, cascades, and the most delicious scenes of Paradise."[49]

THE RISE AND FALL (AND RISE AND FALL) OF THE COLOSSEUM

Locke, Burke, Smith, and Stewart assume that perception is anchored in the real world and that the experiential possibilities opened by imitation (and association) will be narrowed by cultural tradition, common sense, and the innate dispositions of the human body. These constraints, they suppose, are sufficient to ensure that "the economy of our aesthetic-affective life is a natural one."[50] But the emergence of the panorama arguably marks the moment in Western culture when—vying with the edifices of sovereign power and the liturgical spaces of religion—commercial, artificially constructed, secular environments become powerful enough to encroach on the domain of the real. Already in the Colosseum, the multiplication of perceptual environments and the degree to which their phenomena seem to be real suggest, on the one hand, that "the natural *objects of our perceptions are virtual appearances*" and, on the other hand, that "what art and technology do is extend the body's existing regime of natural and acquired artifice, already long in active duty in producing the 'virtual reality' of our everyday lives."[51]

The history of the Colosseum, its rise and fall (and rise and fall), is therefore entwined with the histories of the artist-entrepreneurs (and their financiers) who in a volatile economic and cultural environment reinvent its illusions, forging again and again its perception machines in ways that ensured, albeit only ever temporarily, that they generated interest rather than boredom. First, there were Thomas Hornor, the architect Decimus Burton (1800–81), and their financier Rowland Stephenson (1782–1856). After Hornor followed Stephenson to America in order to avoid bankruptcy, a committee of Hornor's friends managed the Colosseum until 1835, when it was bought by John Braham (1774–1856), the most gifted tenor of his time, and the comedian Frederick Henry Yates (1797–1842). The next in line was David Montague, "the proprietor of the Princess's theatre, Oxford-street,"[52] who after purchasing the Colosseum in 1843 employed William Bradwell (d. 1849), chief set designer at the Covent Garden Theatre, to update its attractions once more.

By the middle of the century, Hornor's original conception had been extended to include Marine Caverns, an African Glen, a "Grand Reception or Banqueting Hall of Mirrors," a theater and concert hall, a Glaciarium (for indoor ice-skating), a Gothic Aviary, an "Exterior Promenade" with ruins intended to remind visitors of "the Temple of Vesta, Temple of Theseus and the Arch of Titus,"[53] a panorama of

"London by Night" (it was lowered each evening over the original panorama), a statue of the Colossus of Rhodes, "The Stalactite Caverns of Adelsberg," a Cyclorama showing the earthquake at Lisbon, and even a model of a silver mine at work. But by 1855, when it was sold to the Colosseum & Arts Company, the Colosseum and its now tawdry worlds, orbiting around a seriously out-of-date view of London, were in terminal decline. In 1862 it was sold for the last time, to the Regent's Park Pantechnicon Company, with John Burns Bryson as the license holder. As Ralph Hyde remarks, Bryson "evidently intended to knock down the building without delay,"[54] but despite these plans it was not until 1875 that the Colosseum was finally demolished.

What, then, is one to make of the Colosseum? If we focus only on Hornor's "Panorama of London," perhaps it could still find a place in conventional histories of the panorama, such as the one set forth by William Galperin in *The Return of the Visible in British Romanticism*, in which conservative spectators are disedified by the panorama's (supposedly) unmediated presentation of the visible world.[55] Conversely, if our attention is captured by the multiplying virtual realities clustering around Hornor's panorama, one might agree with Gillen D'Arcy Wood that it belongs "to a vision of multi-purpose recreational space that prefigured the contemporary Disney-style theme park."[56]

And yet, as I have been suggesting, allocentric and autocentric readings of the panorama understate the significance of the virtual worlds framed by its optical environments *and* by the social display, sympathetic imitation, and social exchange that they fostered, which the Colosseum was designed to intensify and exploit. Seen in this light, the Colosseum looks back to the intense, affective engagement with imagined worlds characteristic of Romanticism, while carrying these phenomena forward into the Victorian period, through the pleasurable experiences it offers of the interimplication of autocentric and allocentric dimensions of experience, and of the consequent malleability of the real. On the one hand, the Colosseum anticipates in broad terms the Victorian enthusiasm for secular enchantments (natural magic),[57] aroused for example by professional magicians, such as Jean Eugène Robert-Houdin (1805–71); John Henry Anderson (1814–74), who was dubbed "The Great Wizard of the North" by Sir Walter Scott (1771–1832); and John Nevil Maskelyne (1839–1917). On the other hand, it can also be seen as a precursor of what at first sight seems a quite different development, represented by the Crystal Palace built by Joseph Paxon (1803–65) to house the Great Exhibition (1850).

Paxton's design for the Crystal Palace was inspired by a conservatory he had earlier built for the sixth Duke of Devonshire

(1790–1858), to cultivate the propensity of the Amazonian water lily (known as *Victoria amazonica* and *Victoria regia*) to flower, which for specimens imported to England had remained stubbornly dormant. In later years, his lily house reappeared in art and polemic as a metaphor for the middle-class home, an environment built to encourage the flowering of bourgeois femininity.[58] Arguably the ambiguously natural and artificial environment opened by the Crystal Palace was designed to cultivate an analogous flowering, this time of the modern, cosmopolitan consumer and his or her protean propensities. As Peter Sloterdijk remarks, the Crystal Palace marks the point from which "a new aesthetics of immersion began its triumphal march through modernity. The psychedelic capitalism of today was already a *fait accompli* in the almost immaterialized, artificially climatized building."[59]

And having now turned our gaze toward the present, we can conclude this list by adding that the Colosseum looks forward to the "marvellous geographies" fashioned by writers such as Arthur Conan Doyle (1859–1930), H. P. Lovecraft (1890–1937), and J. R. R. Tolkien (1892–1973), "which were collectively inhabited and obsessively elaborated by readers,"[60] as well as to online virtual worlds, such as *Second Life*, *World of Warcraft*, and *Eve Online*, which are just as obsessively inhabited and elaborated by players.[61]

As these examples suggest, the Colosseum represents an early experiment in the use of affective space to cultivate propensity, which is understood as a potential that can be exploited. In other words, rather than merely re-presenting the real or communicating a particular ideological message, its virtual realms entice immersants to become cocreators—who through conversation, imitation, and display help shape the virtual worlds that, in the mode of anticipation, have drawn them there. In these developments it is possible already to see, albeit in nascent form, aspects of our own world in which, as Nigel Thrift argues,

> no longer can the value form be restricted to labour at work. It encompasses life, with consumers trained from an early age to participate in the invention of more invention by using all their capabilities, and producers increasingly able to find means of harvesting their potential.[62]

This development, and the (relatively open-ended) affective environments on which it depends, reject the logics of representation/expression, surface/substance, and simple-past/sophisticated-present that have governed histories of the West's visual cultures. Instead they offer terms that describe the space between these poles: imitation, habit, milieu, and (affective) atmosphere, among others. Now, in the early twenty-first century, "affect, imitation-suggestion, and entrancement" can no longer be thought merely romantic and as such "incidental to what the political is and how the political is conducted." Rather, they form the foundation on which new forms of oppression and "more expansive political forms" (both based on a scientifically sophisticated understanding of the malleability and mechanics of propensity) can be built.[63] In this context the Colosseum, along with the panorama phenomenon as a whole, should be considered an important stage in the journey that has brought us to this new-but-old world.

NOTES

1. Nigel Thrift, "Pass It On: Towards a Political Economy of Propensity," *Emotion, Space and Society* 1 (2008): 95. The following argument is a companion piece to the twelfth chapter of my *Multiplying Worlds: Romanticism, Modernity, and the Emergence of Virtual Reality* (Oxford: Oxford University Press, 2011), which places the Colosseum in relation to Thomas Hornor's work as cartographer, land surveyor, landscape gardener, and landscape painter. I would like to thank Liz Wakefield for her help gathering the primary materials on which my argument depends.
2. Jos De Mul, *Romantic Desire in (Post)Modern Art and Philosophy* (Albany: State University of New York Press, 1999), 241.
3. Thrift, "Pass It On" 83.
4. Anon., *The World*, April 11, 1789.
5. Anon., *The Times*, April 24, 1789.
6. Ralph Hyde, *Panoromania! The Art and Entertainment of the "All-Embracing" View* (London: Trefoil, 1988), 62. See also *The Times*, January 10, 1792.
7. Ephraim Hardcastle [William Henry Pyne], ed., *Somerset House Gazette, and Literary Museum; or, Weekly Miscellany of Fine Arts, Antiquities, and Literary Chit Chat*, 2 vols. (London: W. Wetton, 1824), 2:152.
8. Ibid.
9. Francis Bacon, *De Augmentis Scientiarum*, in *The Philosophical Works of Francis Bacon*, ed. John M. Robertson (1905; rpt. Abingdon, UK: Routledge 2011), 447. See also Stephan Oettermann, *The Panorama: History of a Mass Medium* (New York: Zone Books, 1997), 13.
10. Lisa O'Connell, "Dislocating Literature: The Novel and the Gretna Green Romance, 1770–1850," *Novel: A Forum on Fiction* 35 (2001): 5–23.
11. Oettermann, *Panorama* 31, writes of "a new and more 'democratic' perspective."

12. For discussion of this point see Oettermann, *Panorama*,15, and Andrea K. Henderson, *Romanticism and the Painful Pleasures of Modern Life* (Cambridge: Cambridge UP, 2008), 229–30.
13. Brian Massumi, "Envisioning the Virtual," in *The Oxford Handbook of Virtuality*, ed. Mark Grimshaw (Oxford: Oxford UP, 2014), 64. Italics in original.
14. For a discussion of this point in relation to the Colosseum, see Neil Arnott, *Elements of Physics, or Natural Philosophy*, 2 vols. (London: Longman, Rees, Orme, Brown, and Green, 1829), 2:279, portions of which are quoted in *The Mirror of Literature, Amusement, and Instruction: Containing Original Essays*, ed. Reuben Percy, John Timbs, and John Limbird, vol. 14 (London: J. Limbird, 1829), 431.
15. Oettermann, *The Panorama* 32.
16. John Britton, *A Brief Account of the Colosseum* (London: 1829), 4.
17. William Hazlitt, *Sketches of the Principal Picture-Galleries in England* (London: Taylor and Hessey, 1824), 29–30. The Colosseum's *A Description of the Royal Colosseum Re-opened in 1845, Re-embellished in 1848* (twenty-second edition; London: Printed by J. Chisman, 1848) includes a catalogue of 132 pictures and sculptures on display in the Saloon (renamed as the Glyptotheca, or Museum of Sculpture).
18. Anon., *Morning Journal*, February 15, 1830. See also Britton, *Brief Account* 7.
19. Britton, *Brief Account* 3.
20. Anon., "Some Account of the Colosseum," in *The Mirror of Literature, Amusement, and Instruction* 13, no. 352 (January 17, 1829): 35.
21. Britton, *Brief Account* 6.
22. Ibid.
23. Anon., *Morning Journal*, January 19, 1829.
24. Anon., "Some Account of the Colosseum" 34.
25. John Milton, *Paradise Regained*, in *The Poems of John Milton*, ed. John Carey and Alastair Fowler (London: Longmans, 1968), 463, 502–505 (Bk. I, ll. 42 and 710–57).
26. Bruno Latour, *Reassembling the Social: An Introduction to Actor-Network-Theory* (Oxford: Oxford UP, 2005), 39.
27. Bernard Comment, *The Panorama* (London: Reaktion Books, 1999), 86. Italics in original.
28. Otto, *Multiplying Worlds* 30.
29. Peter Otto and Abigail H. Nedeau-Owen, "Humphry Repton: 'View from the House at Repton,'" in *Innovations in Encompassing Large Scenes, Romantic Circles Gallery*, http://www.rc.umd.edu/gallery. Accessed June 10, 2014. See also Jacqueline M. Labbe, *Romantic Visualities: Landscape, Gender, and Romanticism* (Basingstoke, UK: Macmillan; New York: St. Martin's Press, 1998).
30. Otto, *Multiplying Worlds* 30–31.
31. See Oettermann, *Panorama* 31.

32. Anon., "Some Account of the Colosseum" 35.
33. Anon., "The Colosseum," *London Magazine*, February, 1829, 106–7.
34. Ibid., 108.
35. John Locke, *An Essay Concerning Human Understanding*, ed. Alexander Campbell Fraser, 2 vols. (1894; rpt. New York: Dover, 1959), 1:304 (II. xx. 6). For discussion of Locke's views on desire, see Nancy Armstrong and Leonard Tennenhouse, "A Mind for Passion: Locke and Hutcheson on Desire," in *Politics and the Passions, 1500–1850*, ed. Victoria Kahn, Neil Saccamano, and Daniela Coli (Princeton, NJ: Princeton UP, 2006), 131–50.
36. See, for example, Comment, *Panorama* 98–99, and Oettermann, *Panorama* 15.
37. Anon., *A Picturesque Guide to the Regent's Park: with . . . Descriptions of the Colosseum, the Diorama, and the Zoological Gardens . . .* (London: John Limbird, 1829), 32.
38. *The Times*, no. 13347, August 2, 1827, 3. Quoted in Ralph Hyde, *The Regent's Park Colosseum* (London: Ackerman, 1982), 39.
39. Hyde, *Regent's Park Colosseum* 39.
40. Anon., "Some Account of the Colosseum" 36.
41. This scheme was abandoned in December 1828, when Rowland Stephenson, who had provided the funds necessary to build the Colosseum, fled to America after being accused of embezzlement. For contemporary accounts of this debacle, see *The Morning Chronicle*, January 10, 1829; *The Examiner*, January 11, 1829; *The Newcastle Courant*, January 17, 1829; and *The Aberdeen Journal*, June 30, 1830.
42. John Britton and Augustus Pugin, *Illustrations of the Public Buildings of London*, 2 vols. (London: J. Taylor, 1825, 1828), 2:274.
43. Edward J. Diestelkamp, "Fairyland in London: The Conservatories of Decimus Burton," *Country Life*, May 19, 1983, 1342.
44. Anon., *The Times*, January 13, 1829.
45. Anon., "Some Account of the Colosseum" 36.
46. Britton, *Brief Account* 8.
47. *Summer: The Present Fashions*, published and sold by B. Read, 12 Hart Street, Bloomsbury Square, London, c. 1830. The print can be viewed online at http://www.motco.com/imageone.asp?Randompic=90004007. For an account of Read and his printmaking activities, see Ralph Hyde and Valerie Cumming, "The Prints of Benjamin Read, Tailor and Printmaker," *Print Quarterly* 17 (2000): 262–84. As Hyde and Cumming note, the print displays fashionable attire rather than dress appropriate for the occasion (272). I would like to thank Gayle Otto for identifying the plants on display in this design.
48. Jean Starobinski, *Diderot dans l'espace des peintres; suivi de, Le sacrifice en rêve* (Paris: Réunion des musées nationaux, 1991), 29. Quoted in Comment, *Panorama* 98.
49. Britton, *Brief Account* 7.

50. Stephen K. White, *Edmund Burke: Modernity, Politics and Aesthetics* (Thousand Oaks, CA: Sage Publications, 1994), 33.
51. Brian Massumi, "Envisioning the Virtual" 64.
52. *Illustrated London News*, October 14, 1843.
53. Hyde, *Panoramania!* 95.
54. Hyde, *Regent's Park Colosseum* 58.
55. William Galperin, *The Return of the Visible in British Romanticism* (Baltimore: Johns Hopkins UP, 1993), 34–61.
56. Gillen D'Arcy Wood, *The Shock of the Real: Romanticism and Visual Culture, 1760–1860* (Houndmills, UK: Palgrave, 2001), 5.
57. Simon During, *Modern Enchantments: The Cultural Power of Secular Magic* (Cambridge, MA: Harvard UP, 2002).
58. Margaret Flanders Darby, "Joseph Paxton's Water Lily," in *Bourgeois and Aristocratic Cultural Encounters in Garden Art, 1550–1850*, ed. Michel Conan (Washington, DC: Dumbarton Oaks Research Library and Collection, 2002), 255–83.
59. Peter Sloterdijk, *In the World Interior of Capital*, trans. Wieland Hoban (Cambridge: Polity Press, 2013).
60. Michael T. Saler, *As If: Modern Enchantment and the Literary Prehistory of Virtual Reality* (Oxford: Oxford UP, 2012), 6.
61. See, for example, Tom Boellstorff, *Coming of Age in Second Life: An Anthropologist Explores the Virtually Human* (Princeton, NJ: Princeton UP, 2008), and Bo Moore, "Inside the Epic Online Space Battle That Cost Gamers $300,000," *Wired*, February 8, 2014, http://www.wired.com/2014/02/eve-online-battle-of-b-r/?cid=co18382344.
62. Nigel Thrift, *Non-Representational Theory: Space | Politics | Affect* (London: Routledge, 2008), 48.
63. Ibid., 253.

Chapter 9

The Imperial Avatar in the Imagined Landscape: The Virtual Dynamics of the Prince of Wales's Tour of India in 1875–76

Ruth Brimacombe

N.B. Additional images associated with this chapter are housed in the digital annex at www.virtualvictorians.org.

Working on the principle, established by Anne Friedberg, that the virtual realm was "an operable philosophical concept in the late nineteenth century,"[1] this chapter demonstrates how the idea of virtual reenactment lay at the heart of the pictorial representation of the Prince of Wales's tour of India in 1875–76, and was a key factor in this event's popular success in Britain. Taking virtuality, in its modern conception, to be an artificial simulation of reality, this chapter reveals the virtual dimension of the mimetic techniques used to recreate the prince's journey in the pages of the illustrated press. The prince in question was Albert Edward (nicknamed "Bertie"), the eldest son of Queen Victoria and Prince Albert, who subsequently reigned as Edward VII. His tour of India lasted four months in total, while the entire journey took seven—from October 1875 until May 1876. At this time, Queen Victoria's realm included British India, and since 1858 she had used the title "Queen of India." Because the prince officially traveled around India as the "Heir Apparent," his visit became a kind of royal progress. Accordingly, it received an unprecedented level of pictorial press coverage.[2] Four artists, three working for the illustrated press and one for the prince himself, were tasked with the project of visually documenting his entire journey. These so-called Special Artists, or artist-reporters, accompanied him everywhere, sending a steady stream of the resulting imagery back to Britain via the pages of *The Illustrated London News* and *The Graphic*. On their return, two of the main artist-reporters involved in the tour participated in special exhibitions in London, showcasing the more finished versions of the sketches they had produced for the press.[3] This ensured that the British public's imaginative engagement with the prince's Indian tour was intensely sustained for about a year. While the art critics reviewing these exhibitions generally struggled to evaluate the works artistically, they admired the artist-reporters' remarkable ability to convey a sense of vicarious experience. Here is the reviewer for *The Builder* commenting on William Simpson's *India "Special"* exhibition that opened in June 1876:

> Looking over these numerous and able sketches, . . . one cannot but reflect on the immense value of such attendance upon a semi-royal progress of this kind. But for the efforts of these gentlemen, wielding the pen of the ready writer and the pencil of the ready draughtsman, the tour which has been followed with such interest by millions of people would have been a mere matter of hearsay, of no interest at all, except in any ultimate results, to those at home. As it is, between the pen and the pencil we almost follow the whole thing, and feel as if we had been there.[4]

This acknowledged capacity of pictorial reportage to generate a collective sense of virtual presence, and the wider implications of this phenomenon, are the focus of investigation in this study.

The "numerous and able sketches" that comprise the artistic record of the Prince of Wales's Indian tour are a particularly salient example of a pioneering type of documentary art that came to prominence in the nineteenth century as a result of the establishment of *The Illustrated London News* in 1842. This publication took advantage of technological advances, made at the end of the last century, that allowed for the printing of wood engravings alongside text; it consolidated its success by using the innovation, inaugurated in response to the Crimean War of 1854, of sending an artist to the scene of newsworthy events to provide a series of firsthand visual accounts.[5] The groundbreaking significance of this combination of cultural activities—the opening of new channels of media communication in which a sequence of illustrative images linked to the same topic appeared alongside corresponding text, and the forming of a new category of artistic profession that relied on the artist's reportorial ability to accurately describe an actual event—was recognized in its day but has since fallen into obscurity.

Arguably, one of the main reasons for the broad neglect of pictorial reportage has been the relative inaccessibility of the publications in which the images were reproduced—with copies of the weekly newspapers being collected and bound biannually into heavy, unwieldy library tomes—as well as the seemingly ephemeral nature of their original material form, the sheer number of images involved, and the lack, then and now, of a critical framework to adequately define them. However, an analogy used by the correspondent George Augustus Sala in relation to *The Illustrated London News*'s then-imminent coverage of the prince's Indian tour, in which he equates the viewing of the forthcoming reports with the watching of an "artificial moving panorama" at the Egyptian Hall, shows how these long sequences of reportorial images are best evaluated. In effect, one should take the approach identified by Sala when he encourages those

> who are courteous enough to peruse this performance to imagine, for the nonce, that they are seated in one of those comfortable stalls which I spoke of anon, and that they are about to behold the unrolling of a panorama painted by the very best scenic artists—I mean draughtsmen on wood and engravers—of the day. Never mind how many thousand strong the audience may be: the proprietors of the ILLUSTRATED LONDON NEWS will find room for them all.[6]

By accepting that this genre of graphic reportage, usually generated in response to an episodic event such as a royal journey or a military campaign, belongs to the realm of popular entertainment rather than the canon of fine art, and by seeing the long series of images it produces as akin to the panorama, one can locate it as an art form within two other related traditions: that of the travelogue and that of protocinematic spectatorship. This perceptual realignment also brings illustrated journalism into line with the work of scholars in the emerging field of Victorian media studies, who have already established the virtual dynamics of these associated developments.[7] In her enlightening study, *Shivers Down Your Spine: Cinema, Museums, and the Immersive View*, Alison Griffiths touches on the "intermedial" link between the panorama and the illustrated newspaper, noting how, in its reconstruction of scenes inspired by topical events, the former was seen to function in the manner of the latter.[8] My study addresses, for the first time, how this parallel also operated in reverse. In so doing, it positions the category of pictorial reportage firmly within the discourse surrounding the phenomenon of "virtual travel"—that is, the vogue for recreating actual or imaginary journeys through the medium of illustrated representations, which began in the eighteenth century but proliferated in the nineteenth with the growth in mass media industries.[9] The visual reports of the prince's Indian journey were undoubtedly examples of virtual travel. In Sala's introductory feature, the correspondent not only conjures up the image of the "armchair traveler," but also refers to a prevalent claim associated with this figure.[10] Citing the paradoxical view that an exercise in virtual travel could be more edifying than an actual voyage, Sala declares:

> We may learn a great deal from an artificial moving panorama when we are seated in a comfortable arm-chair and the panorama glides gently before us. On the other hand, we are apt to derive but very little instruction from a natural panorama, which is stationary, while we dash past it in express-trains or rapid steam-ships. To find those who may in the greatest measure profit by the Royal trip to Hindostan I venture to look at home. The graphic and animated description of the Prince's tour . . . should awaken in the minds of the public at large a lively and a lasting interest in India and all appertaining to it.[11]

In using this terminology, Sala echoes the ambition voiced by *The Illustrated London News* at the time of its founding "to keep before the eye of the world a living and moving panorama of all its activities and instances."[12] On one level, these examples of the illustrated journal explicitly aligning its effects with those of the moving panorama are simply emblematic of the way in which the experience of panoramic viewing turned into a standard frame of reference for comprehending situations of modern spectatorship.[13] However, in her insightful study of the formative role of virtual travel in the conception of Victorian realism, Alison Byerly takes this idea further and explores how the panorama developed into a perceptual tool for imaginary journeying, which in turn shaped the practice of envisagement in various literary forms.[14] As part of this enquiry, she notes that by using the panorama as a metaphor, *The Illustrated London News* was

> invoking a power to synthesize, reflect, and channel the disparate forms and events of modern life into a unified representational stream that would carry the reader/viewer along with it. This synthesis of word and image would allow the public to have "under their glance, and within their grasp, the very form and presence of events as they transpire, in all their substantial reality."[15]

What Byerly's observation and my study confirm is the consciousness with which the illustrated newspaper strove to develop these virtual properties and how intrinsic the notion of virtuality was to the publication's core purpose of visualizing the news. Taking the view that the illustrated newspaper was successful in its endeavor, this chapter further considers the reifying qualities of the art on which the newspaper relied to create the illusion of reproducing the "substantial reality" of the prince's tour.

Too often, when this kind of journalistic imagery is discussed, it is in isolation from any consideration of the technical skill or intellectual agenda of the artist responsible for its production. Yet what makes the prince's tour such a revealing case study is the prominence of the artists involved in the event's documentation and the sheer breadth of the entire body of artwork relating to the tour that is available for study. Though it is widely dispersed, the archive in aggregate includes a large residuum of preparatory pencil sketches, illustrators' sketches drawn in India ink with white highlights, black-and-white wood engravings as they appeared in the press, and the final, colorful watercolor versions of the scenes exhibited at a later stage. Far from anonymous, William Simpson, the artist commissioned on this occasion by *The Illustrated London News*, enjoyed considerable fame as "Crimean Simpson." Trained initially as a lithographer, he gained this soubriquet because he was one of the first artists sent to the theater of war to provide visual reports of the military events then preoccupying the British public's imagination. In a radical move, Colnaghi's, one of the leading firms of printsellers, had employed him to supply images direct from the front, while, on this occasion, *The Illustrated London News* called on the services of the French artist Constantin Guys. Having moved to the staff of *The Illustrated London News* in 1866, Simpson, as their preeminent artist-reporter, was widely acclaimed as "the doyen" of this new artistic profession by 1875.[16] Toward the end of his career, Simpson wrote several self-reflexive, autobiographical accounts looking back at the unique qualities of his occupation as an artist-reporter, which revealed his own appreciation of the substantializing character of his work. In one of the draft versions of a feature on the "Special Artist" he wrote for an issue marking the Queen's jubilee of 1892, he commented:

> Now the artist sees what takes place, and his work is immediately given forth to the world, so that its accuracy can be tested even by the actors in the historical event. Illustrated Journalism represents this new Avatar of Art, and, although its origin may be traced further back, its real beginning and development came all within the time since Queen Victoria began her rule.[17]

His use of the term "Avatar of Art" is salient. Knowledgeable about Indian theology as a result of long periods of travel in India and his own scholarly interests, Simpson alludes to the Hindu idea that divine (and therefore abstract) forces can be incarnated in pictorial or sculptural form to articulate his understanding of the reifying effect of the journalistic art that he practiced.[18]

This notion of pictorial journalism having the power to manifest history in visible form was common in the writings of his reportorial colleagues. In compiling the first history of the profession, *The Pictorial Press: Its Origins and Purpose* (1885), Simpson's contemporary Mason Jackson celebrates the way in which "treasures of truth that would have lain hid in Time's tomb" had become available as a result of "the enduring and resuscitating powers of art—the eternal register of the pencil giving life and vigour and palpability to the confirming details of the pen."[19] How successful the artist-reporters were in meeting this requirement to revivify current affairs for a present and future audience was the tacit scale on which their abilities were judged. William Simpson excelled, while Walter Charles Horsley and Herbert Johnson, who were commissioned to report on the prince's tour by the rival journal *The Graphic*, were seen as promising but as yet unproven in the field.[20] The fourth artist following the tour, Sydney Prior Hall, had joined the team of staff artists at *The Graphic* soon after it formed in 1869 and had quickly made a name for himself reporting on the Franco-Prussian war of 1870–71. Hall would have a long and illustrious career reporting on royal events for *The Graphic*, and the critic Lewis Lusk in his later assessment of Hall's journalistic oeuvre would praise his skill for providing "vivid presentments of the scenes enacted."[21] However, on the occasion of the Indian tour, Hall put his reportorial skills at the disposal of the Prince of Wales and traveled with the royal entourage. While Hall's works were therefore not part of the initial "unified representational stream" that appeared in the pages of the pictorial press, the imagery he produced for the prince, which was subsequently exhibited to the public and used to illustrate William Howard Russell's published account of the visit, still relied on journalistic techniques and thus contributed to the larger raft of narrative imagery generated by the tour.[22] The same is true of the original sketches that Simpson put on display after the conclusion of the tour, which were also subsequently published in photographic form.[23] As indicated by the *Builder* reviewer's response to Simpson's *India "Special"* exhibition, quoted at the start of this chapter, the unifying characteristic of all the graphic imagery created around the subject of the tour was its ability to facilitate the sensation of virtual travel and to build an empathetic sense of bodily presence—in effect, to simulate an imaginary model of the visit that allowed the reader/viewer to follow the prince on his journey and thereby to enjoy a virtual encounter with India.

Art's role in the formulation of an imagined India is well established. Several scholars exploring the visual representation of India by British artists and photographers have demonstrated how, since the early days of the East India Company, the dissemination of printed images of India—in tandem with written memoirs and "picturesque travel-books"—had helped to construct an imagined landscape of the subcontinent in the mind of the British public.[24] The pictorial reportage of the prince's journey in the illustrated press exemplified a development that both formalized this tradition and extended it beyond the elite to a broader middle-class readership. As Sala proclaimed, no matter "how many thousand strong the audience," the illustrated press, with its substantial circulation figures, aspired to allow all their readers access to this exercise in imaginative geography.[25] The earlier publication of works by British traveling artists, including William Hodges and Thomas and William Daniell, had instituted a visual canon of iconic Indian sites that included famous locations, such as the rock-cut caves of Elephanta.[26] Utilizing a technique also employed by the panorama designers, both *The Illustrated London News* and *The Graphic* deliberately structured the fresh imagery being supplied by the artist-reporters around reiterations of these classic views of India and around selected examples from a store of typical views of Indian life and customs.[27] In this manner, the illustrated journals purposely evoked the imagined topography of the country, cumulatively established by European observers of India since the eighteenth century, as the virtual arena for the prince's actual journey.

In her discussion of the panoramas that recreated celebrated river journeys, such as Thomas Banvard's presentation of a trip along the Mississippi, Griffiths makes an important distinction between the panoramas that reconstructed a generic, touristic experience of traveling along a well-known riverscape and those that recalled an individual's singular experience of a specific journey.[28] A similar analysis of the imagery comprising the visual reports of the prince's Indian tour shows that the artist-reporters carefully intermingled these two modes. Take, for instance, William Simpson's inclusion of a picturesque image such as *The Royal Visit to India: The Ganges* (see annex), in which he openly invites the viewer to revel in a moment of unalloyed virtual tourism. Here, adhering to the pattern set by his eighteenth-century artistic predecessors, whose work he had studied at Indian House before embarking on his first artistic odyssey around India in 1859, Simpson offers a timeless Indian vista; he shows a ghat (a landing point) and a group of menacing crocodiles on the banks of the river, placed as if they were thrillingly close to the viewer.[29] This image, in its style, also indicates Simpson's allegiance to the methods of his artistic contemporaries, the topographical artists David Roberts and Edward Lear, who similarly specialized in depicting exotic vistas that encouraged the viewer to vicariously enjoy a pictured landscape scene made from "on the spot" studies.[30]

Conversely, works such as *Illuminations at Benares* (see annex) mark the point at which Simpson and his fellow artist-reporters begin to depart from these established aesthetic conventions and do something qualitatively different. Note how in this case the artist includes within the composition the flotilla of barges conveying the royal party and the precise inscription "Illuminations at Benares, 5 Jan 1876." These factual references mean that while Simpson's Ganges image could be abstracted out of the "unified representational stream" and enjoyed in isolation as a typical view of India, his depiction of the scene in Benares, though equally evocative, is firmly grounded in the terms of an actual event. The artist has particularized the view to that experienced by the Prince of Wales and his entourage at a certain time and place.[31] This shift in focus reflects the emergence of a new journalistic idiom for an art that is concerned with the hyperaccurate representation of factual narratives—and the growth of a new type of travel imagery, driven by a developing interest not just in the places an individual has seen, but also in the sensations and physical conditions of actual travel.[32]

Accordingly, great emphasis was placed on the experiential effect of the work produced by these artist-reporters. In order to capture a sense of the prince's journey in its entirety, their pictorial reports included scene-setting images that charted his physical situation from the moment he stepped on board his boat, the *Serapis*, at the start of his voyage until his disembarkation back in London. As a result, the body of imagery ranged from prosaic views of the Prince of Wales walking up boat gangways and reading correspondence in his tent in the jungle to grandly romantic set-piece scenes of him presiding over ceremonial occasions. Simpson's depiction of him in *The Prince of Wales Mounting his Elephant at the Old Palace of Lushkur, Gwalior* (Figure 9.1) neatly demonstrates this tendency to highlight the incidental as well as the symbolic.

Interjected between the artist's representations of the prince's spectacular entries on the back of a caparisoned elephant were "behind the

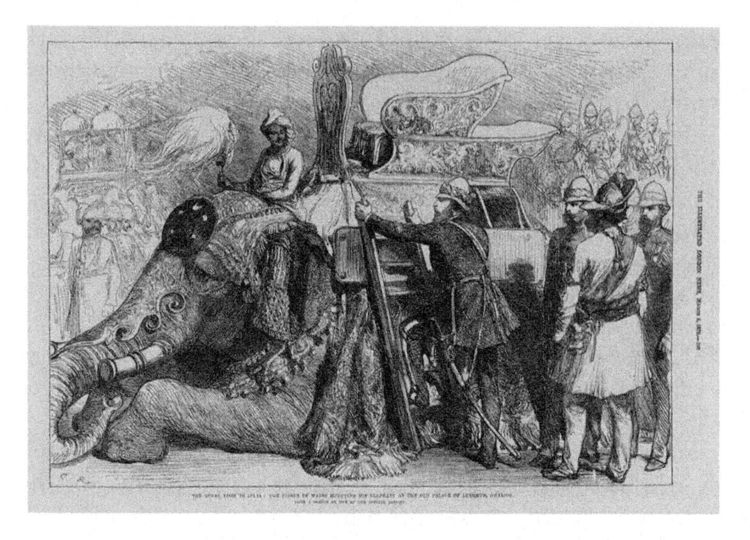

Figure 9.1 After William Simpson, *The Royal Visit to India: The Prince of Wales Mounting his Elephant at the Old Palace of Lushkur, Gwalior*, wood engraving, *The Illustrated London News*, March 4, 1876, 237.

scenes" images, such as this one; it shows the prince poised with one foot on a ladder, engaged in the more mundane task of clambering up into the beautifully decorated howdah on the elephant's back.

Building on the illustrated journals' practice of interleaving their new visual reports with standard views of India, the artist-reporters also depicted numerous scenes in which the Prince of Wales appeared to be figuratively embedded *within* this imagined Indian landscape. So, for instance, in Simpson's striking view, *The Prince of Wales Lunching in the Caves of Elephanta, 11 December 1875* (see annex), the prince and his party are seen inside the main cave, seated in front of the Trimurti, a triple-headed statue of the Hindu deity Shiva, enjoying a very English lunch party within the sacred confines of one of India's most famous archaeological sites. In light of Edward Said's well-known arguments concerning the malign effects of Orientalism—that is, the imagining of the East by the West, in which the imperial regime controlled the perception of the cultures it subjugated—the implications of such an act of imperial insensitivity are now startlingly evident.[33] Commentators have noted how the sense of "mastery" inherent in the panoramic spectacle, because of the seeming totality of the vision it offered, lent itself, in perceptual terms, to this form of visual appropriation.[34] Illustrations such as this extraordinary scene in the Elephanta caves indicate how the virtual forums opened up by the pictorial press were equally implicated in this imperial practice.[35]

Ultimately, the pictorial reenactment of the prince's tour enabled by the illustrated journals was contingent upon the mode of production of their "illustrative art" and the format through which this imagery was visually communicated. Extant copies of the pencil drawings William Simpson made while observing the events in India, such as the one inscribed "Figures—Laying Foundation Stone with Masonic honours by the Prince of Wales, of New Docks. Bombay. 11th Nov 1875" (see annex), provide a rare glimpse into the working methods of a professional artist-reporter. These show how Simpson first made quick shorthand notes recording details of the profiles, costumes, and physical appearances of key participants in the event. He then used these records as an aide memoire for the production of an "illustrator's sketch"—a more-polished draft composition in pencil, ink, and wash—that provided his subjective impression of the most definitive aspect of the witnessed event. A surviving example showing Simpson's piece *H. R. H. The Prince of Wales at the School Children's Fete Bombay, 10th November 1875* (see annex) demonstrates how, for the sake of speed, the artist concentrated on detailing the look of the main characters in the scene, while only sketchily suggesting the forms and positions of the surrounding crowd. Typically, as in this example, Simpson then included descriptive annotations to guide the staff draftsman back in London responsible for translating his image into a press illustration. A standard four weeks after the event, the wood engraving of this view of the prince being garlanded with flowers by a Parsee schoolgirl appeared in *The Illustrated London News* on December 11, 1875 (see annex). A comparison of the two stages of the composition reveals how the engraver has retained and sharpened the essential features emphasized by Simpson—thereby preserving the indexical link to the actual event—while improvising the background detail given in the engraved version.

The front pages of the periodicals were dominated by single headline images from the prince's tour for a period of approximately eight months. Typically, the interior of the journal would include a double-page illustration and perhaps three or four further images set within several columns of corresponding, descriptive text. Numerous special supplements with additional images and long written accounts were also produced. This deliberate interplay between text and image creates a sense of animation, as confirmed by the critic reviewing Simpson's exhibition for *The Builder*, who marvels that "between the pen and the pencil we almost follow the whole thing, and feel as if we had been there."[36]

So, for instance, Simpson's impression in his piece *Prince of Wales on Board the State Barge of the Maharajah at Benares* (see annex) is intensified by the vivid accompanying description of the hull being "painted a grey green" and decorated with "water-plants, fish and aquatic birds," with a canopy "covered with gold fringe" and "seats beneath of blue velvet."[37] Notably, Sala emphasizes that it is this combination of the "graphic and animated description" of the prince's tour that should "awaken in the minds of the public" a sustained interest in India.[38] As with the serial reading model of the novel, similarly published in the weekly editions of the periodical press, the continuity of subject matter promotes—despite interruption—a protracted emotional and mnemonic engagement with the unfolding narrative. And the interlinked written detail encourages the viewer to mentally enliven the sequence of documentary images in his or her imagination—an effect now identifiable as a form of virtuality.

The continuities between the verisimilitude sought by the artists, authors, and impresarios of the nineteenth century and the construction of virtual environments through digital means in today's world have become apparent to scholars working on various cultural forms—including those analyzing the realist novel, popular spectacles (such as the panorama and diorama), and the travelogue.[39] Yet this thinking has never before been applied in depth to the serial illustrations of the pictorial press.[40] However, the widespread nature of this line of thought indicates the way in which the digital advancements of the current age have provided the conceptual apparatus needed to retrospectively comprehend the full impact of the mimetic developments of the nineteenth century, and to see in them the functional and psychological antecedents for the digital construction of "artificial reality" in the present.

Meanwhile, in purely practical terms, the advent of the digital camera has been the critical factor that has enabled a reappraisal of the long streams of serial imagery in the pictorial press as a discrete body of work. By making digital copies of the widely dispersed individual images and collecting them within a database, scholars can reconstitute the archive of textual and pictorial representations associated with a particular news story, such as the Prince of Wales's visit to India. And through this recovery of the archive, the "unified representational stream" of images becomes evident and accessible once more as a distinct and coherent body of visual data. Having been abstracted from the pages of the pictorial press, the digital versions can be stored in a manageable format that allows for analysis of the patterning of the broad spectrum of images and of their sequential narrative structure and flow. In this way, digital technology also provides the mechanism needed to recapture for a contemporary viewer a sense of the compelling initial impact of this graphic reportage upon its original audience, by giving visual substance once more to the idea of the serial illustrations unfurling in the manner of a moving panorama. However, while the virtual dynamics of the pictorial press are thus made more intelligible to the current scholar, it is important to emphasize that the digital forum does not necessarily bring out virtual characteristics that have remained latent within the periodical press. It is clear that Victorians were themselves already cognizant of these virtual applications, with, for instance, Sala explicitly describing the publication of his introduction to the prince's visit as being "the first stone of an Imperial monument . . . virtually laid in the memory of the British nation."[41]

Looking back, a set of common characteristics now links the realist novel, the travelogue, and panoramic scenography with the cyber realms constructed by the artificial reality technology of the present day. Certain virtual traits mentally transport the viewer or reader into an imagined time and place and provide a facsimile experience by evoking a vicarious sense of physical presence. These hallmarks of virtuality, which involve creating the sensation of an immersive experience, the illusion of mobility, and the development of affective qualities, are just as evident in the series of images produced by the artist-reporters.[42] Given illustrated journalism's participation in the practice of virtual travel, the power of the pictorial reenactment of the prince's journey lies entirely in the artist's ability to convincingly restage the exact circumstances of the tour in the public's imagination.

For example, Simpson's view of the prince's hunting party *Crossing a Nullah in the Terai* (see annex) is presented from the viewpoint of someone looking down upon the passing elephants but from within the group. It demonstrates the artist's skill in replicating the palpable environment of the hunting terrain, with its deep ravines and clumps of dense grass cover. Both this image and his representation of the royal party lunching in the caves of Elephanta exemplify the way in which Simpson elicits from the viewer a strong empathetic sense of being physically located within the depicted space. His work, and that of the other pictorial journalists, also displays the motile quality that is necessary to establish virtuality convincingly.[43] Take, for instance, the implicit sense of the elephants' heavy gait in Simpson's view of the prince's party pushing their way through the undergrowth of the

Terai jungle toward the horizon line of the composition, the tensed figure of the prince as he prepares to ascend his caparisoned elephant, or the forward motion of the state barge moving along the river in Benares.

Since the virtual realm as a concept hinges on this notion of simulated presence, the virtual qualifications of pictorial reportage become clear. The capacity for conveying a sense of immediacy or "being present" is both a paramount feature of the artist-reporter's oeuvre and the recognized common denominator of realist art forms identified as operating in a virtual manner.[44] Sydney Prior Hall was considered particularly adept at relaying a bodily sense of the events he attended. In *The Grand Chapter of the Star of India at Calcutta, 1 January 1876* (see annex), intended to chronicle the high point of the prince's set-piece ceremonial activities in India, Hall reproduced his firsthand experience of the observed scene by capturing the distinctive gestures of the watching crowd (including the whispered conversations between the British officers and the dozing figure of one of the Indian princes) and by modeling the dense, heady atmosphere of the large, packed marquee with its clutter of ceremonial paraphernalia. This replicative ability marks out Hall as another exponent of the "art of embodiment"—an alternative strand of realism identified at the start of this century by the theorist Michael Fried in relation to the work of the German realist painter Adolf Menzel,[45] a connection in artistic practice first noted by the critic Lewis Lusk at the turn of the twentieth century.[46] As a result, Fried's writings on this theme now offer a useful hypothesis that helps to elucidate the virtual dynamics of journalistic art.[47]

As indicated, Hall and Simpson, and to varying degrees the other "Special Artists," relied for imaginative effect, like Menzel, on evoking the viewer's empathetic response to the intuited substance of the depicted scene—evidently with some success, given the reviewer's opinion on viewing Simpson's exhibition that it made the visitor "feel as if we had been there."[48] Positioning Menzel in opposition to John Ruskin's realist school of thought, with its focus on "ocular veracity," Fried describes the artist's use of what he terms "the corporeal imagination," by which he means the ability on the part of Menzel to elicit from the viewer an innate, sensory knowledge of the represented object and its material form.[49] Developing the idea that Menzel's works are shaped by this "affective modality," Fried sets up the notion of the artist as a proxy viewer. He identifies how Menzel's images are structured around "a specific angle of vision" and contain "spatial cues" that capture the artist's "situatedness." This artistic format is apparent in the artist-reporters' retelling of their lived experience of the prince's tour.[50] As with examples by Menzel, works such as Sydney Prior Hall's piece *The Grand Chapter of the Star of India* or Simpson's drawing *The Prince of Wales Lunching in the Caves of Elephanta* use the perspective of the artist's individual line of sight and similarly mimic the "actual physical movements—the actual inclinations of the head and neck—with which an embodied viewer would have perceived the original scene."[51] This ability to simulate personal experience, equally evident in the work of Constantin Guys, Simpson's predecessor at *The Illustrated London News*, attracted the attention of the French critic and theorist Charles Baudelaire, who marveled at the power of the apperception he received from Guys's works as a result:

> Under the spur of so forceful a prompting, the spectator's imagination receives a clear-cut image of the impression produced by the external world upon the mind of Monsieur G. The spectator becomes the translator, so to speak, of a translation which is always clear and thrilling.[52]

Recognizing the parallels between the digital media of today and the verisimilitude of the past cues a reconsideration of the role of the avatar in this context. An obvious corollary of the ideation of "virtual travel," and the ability to "magically transport" a reader or viewer into a "real-seeming world," surfaces with the notion of the key figure in the work of travel art (or main protagonist in the case of the novel) functioning as a substitute visitor—in effect as an avatar, in the modern sense of the word as an online proxy persona that allows a participant to notionally move through a digitally created virtual world.[53] The art historian Leonard Bell, with reference to the traveling artist Augustus Earle, who often represented himself within his images, notes how "the figure of Earle functions as a determinant of how viewers see or experience what is otherwise depicted, as a figure who guides the narrative."[54] Crucially, the portrayal of the Prince of Wales in the illustrated accounts of his Indian journey works in exactly this manner; as a result, encouraged by his celebrity at home, and by the prevalence of his domestic image, the Victorian public shared his touristic experience of India by mentally following his familiar form through the depicted landscape of the subcontinent.[55] William Simpson's depiction of him in *The Prince of Wales on Board the State Barge of the Maharajah of Benares* represents a literal realization of this effect. The strong focal point of the drawing is the recognizable figure of the prince seated on the lotus throne in the prow of the barge, with the splendid decoration of the two rearing horses behind him.

The perspective of the view is deliberately partial. One of his aides directs the prince's attention to the riverbank, seen indistinctly in the distance, but the image concentrates on providing a close-up view of the prince in the act of experiencing the occasion. This observational motif predominates throughout the pictorial reports. There are a multitude of examples, too numerous to illustrate here, that show the prince watching a spectacle, such as the Perehara Festival in Kandy, or looking at a celebrated view, like Walter Charles Horsley's image of the prince at "Sensation Rock," precariously riding on the foot plate of a locomotive engine, staring down at the dramatic landscape of the famous ghat passing beneath his feet.[56]

Yet, clearly, the implicit figure of the "embodied artist" also functions as an avatar for the viewer's imagination, becoming the vehicle for their "feats of imaginative projection."[57] Simpson would himself articulate this concept, describing the sense of weighty responsibility he felt on attending one of the Prince of Wales's ceremonies in India: "Through my eyes the British public would look on the ceremony—to which it might have been added that eyes even yet unborn would also see it."[58] Significantly, the success of the panorama as a mode of spectacle similarly hinges on this concept, with commentators now recognizing that its essential appeal lay in the invitation "to share communion with the artist" and the way in which "through the repetition of place, time and a 'seeing body,'" the visitor's "vision is aligned with that of the artist."[59]

Appropriately, the best and most emphatic visual explication of this phenomenon is provided by Simpson, whose illustration *The Royal Visit to India: The Phul-Bagh, or Flower-Garden, Gwalior* (Figure 9.2) allows the viewer an almost visceral, dizzying sensation of experiencing the vertiginous drop down to the landscape depicted below. Viewers can see this park in Gwalior spreading out in front of them as if they were looking through the artist's own eyes, from his elevated vantage point on a balcony, in what would in today's media lexicon be referred to as "real time."

Of course, any such claim about the seemingly "unmediated" quality of the documentary realism achieved by artist-reporters is problematic, given our understanding of the flawed subjectivity of the image maker. Historians have charted how the unquestioning acceptance of the ocular testimony of art in the eighteenth and early nineteenth centuries was replaced by an increasing awareness that such apparently authentic and realistic images could bear false witness; this is the "mimetic fallacy" with which we still grapple today. This new awareness reportedly led to a shift in preference for images achieved through

Figure 9.2 After William Simpson, *The Royal Visit to India: The Phul-Bagh, or Flower-Garden, Gwalior*, wood engraving, *The Illustrated London News*, March 4, 1876, 236. © National Portrait Gallery, London

mechanical means—such as the camera—and then to a dawning realization in the twentieth century that even these photographic images were equally fallible.[60] While acknowledging the role of the illustrated press as an important new provider of visual information about the British Empire in the mid-nineteenth century, historians of photography and topographical art tend to promote the idea that the camera quickly superseded the mimetic faculty of the artist's eye and offered a groundbreaking form of imperial vision. Along with Susan Sontag, who famously declares that "painting never had so imperial a scope," these scholars maintain that photography inexorably progressed to become the dominant force in the visual communication of the British Empire from the mid-nineteenth century onward—and that the kind of realism engendered by the camera fundamentally altered the nature of that communication.[61] However, this view does not fully account for the sophisticated virtual effects achieved by the artist-reporters who covered the Prince of Wales's Indian journey in 1875–76; it seems that the contribution of illustrated journalism to the visual history of the nineteenth century needs to be fundamentally reassessed.

Such a reassessment must account for the role photography played in the virtual agency of the pictorial reports. Bourne & Shepherd, "the best known of Indian photographers," reportedly "deputed the chief of their staff" to accompany the Prince of Wales around India.

In a widely publicized announcement that signaled the novelty of the appointment, the firm proclaimed, using a now-familiar panoramic analogy: "This 'photo-special' will be assisted by a large number of skilled native photographers, who hope in concert to produce a perfect panorama of the Royal Progress through Hindustan."[62] Unlike the artist-reporters, this "photo-special" remains an anonymous presence in the narrative of the tour that emerged, although it is believed that the now-celebrated Indian photographer Lala Deen Dayal was responsible for supplying some of the photographic images marking the prince's time in Jaipur.[63] On his return to Britain, the Prince of Wales purchased six massive souvenir albums from Bourne & Shepherd full of photographs relating to the tour, and the firm also produced a smaller single-album version available for sale to the general public.[64] Mirroring the commemorative formula adopted by the illustrated press, Bourne & Shepherd offered these albums as a bespoke service, whereby the purchaser was able to supplement their selection of the fresh images taken during the prince's visit with stock images from the firm's archive relating to the locations visited in the course of his tour.[65] A couple of photographs were engraved for reproduction in the illustrated press, including one, for instance—featuring the prince posing beside his elephant during a hunting trip in the Nepalese Terai—that appeared in *The Graphic* on June 3, 1876. However, the slow production and dissemination of the souvenir albums means that the images they contained could only ever have played an ancillary role in the documentation of the prince's tour in 1875–76. Thus, contrary to the claims concerning the evolution brought about by photography in the realm of "imaginative geography" and the novelty of the stereoscope's ability to give the viewer a sense of "virtual immersion in the field of vision," the evidence presented in this chapter indicates that, conceptually, it was the artist-reporters who engineered these developments.[66]

A photograph such as *Arrival of the Prince of Wales at Bombay* (see annex)—although highly evocative of the occasion, its location, and the watching crowds—reveals the limits of the type of photographic "realism" offered by the camera at this point. One of a relatively small number of actual scenes provided by Bourne & Shepherd, this photograph is impressive in scale, capturing a bird's-eye view of the procession of carriages that took the prince and his party into Bombay. But it displays the "aura of suspended animation" characteristic of photographic images from this period.[67] Constrained by long exposure times, the photographer was only able to produce static, factual images; in contrast, artist-reporters could respond to the unpremeditated actions of the tour and convey movement. As a result, it is the artists' reportorial imagery that drives the visual narrative of the visit and its appearance of spontaneous immediacy, while photography could only support the imaginative exercise by offering an intensification of the attendant detail. Accordingly, the narrative iconography devised by the artist-reporters would set the blueprint for the documenting of princely tours of India—in film as well as photography—well into the next century.

The most critical point is that their contemporaries already found the work of the artist-reporters capable of providing a convincing equivalent to photographic "realism." Referring to Simpson's dramatic representation of a military review, Archibald Forbes—the special correspondent working for *The Daily News*, who had been present at the event—admired the "strange power of the trained artist's eye" that allowed Simpson to turn the "confusedly blended impression of the galloping horsemen" into an accurately detailed sketch in which "you may see all this, and share for yourself in the 'lurid whirlwind.'" For Forbes, the capacity for precise recreation transcended the category of "impression."[68] Similarly, Harry Barnett, explaining the reportorial process for *The Magazine of Art* in 1883, described how the special artist first prepared "a variety of sketches of the broad general aspect of the scene" as "memoranda" in order to build a mental layout. This was certainly Simpson's working method, judging by the sketches made during his preliminary observations. However, what distinguishes the artist-reporter's talent is the subsequent moment when, according to Barnett, they experience a phenomenon whereby

> the scene resolves itself into a true picture, which remains for a minute, perhaps and then melts. For that supreme moment the Special Artist watches with all his faculties; it bursts suddenly upon him, and instantly he must select and record its essentials. The teeming details, which (though he sketch never so rapidly) he cannot indicate with his pencil, he must photograph in his memory.[69]

Barnett's description, along with the remarks of Forbes and Baudelaire (who refers to "all good and true draughtsmen [drawing] from the image imprinted on their brain, and not from nature"), points to the conclusion that, as an artistic breed, the artist-reporters were in possession of an eidetic faculty that allowed them to recreate the prince's journey extremely persuasively in the pages of the illustrated press.[70] Usually associated with psychological studies carried out on a child's imagination and memory, eidetic power is attributed to someone who has the ability to retain an image after it is no longer actually in front of them. What they have in mind is not simply a lingering afterimage or the projection of a visual memory, but a vividly exact residual image that they still actively see.[71] Certainly, an eidetic aptitude would account for the potency of the artist-reporters representations and would explain how their contemporaries, including those present at the events described, could perceive their work as being quasiphotographic in nature. In any case, as the preparatory stages of Simpson's work demonstrate, the sketches of the artist-reporter, made as acts of recordkeeping rather than acts of invention, preserved a strong indexical link to the facts and appearance of the actual event, in the manner associated with the workings of the camera. Most importantly, Victorian audiences perceived and remarked on this indexical link. Proud of their eyewitness status, the artist-reporters advertised their adherence to a strict code of empirical ethics, and their work was celebrated for achieving an uncanny level of verisimilitude—an effect now explicable in terms of today's conception of the virtual. The "illustrative art" of the "Special Artist" was conceived in the strong belief that it would be of value to "historians of the future," but it has taken the conceptual advances and lexicon associated with the digitally created "virtual worlds" of the present age to provide the frame of reference needed to finally make sense of this potential.[72]

I have shown how the visual medium of pictorial journalism gave its Victorian audience a powerful virtual encounter with India through the reports of the Prince of Wales's journey, and I conclude by underscoring the critical significance of this effect. The contemporary understanding of the construction of an artificial reality once again proves helpful. Scholars have noted how the illusion of virtual travel created by the panorama or travelogue was contingent on manufacturing a convincing mimetic space that, though reliant on components of the actual, was essentially liminal in character. In such a space, in Griffiths's words, "fiction and fact, absence and presence, now and then" could seamlessly meld in the course of a reenactment that was presented as being entirely truthful.[73] The conceptual power of this virtual phenomenon, combined with the eyewitness status of the artist-reporter's work, is the reason that the documentary imagery of the prince's visit, directly communicated to the imagination of the British public through the pages of the illustrated press, proved to be so compelling—and the reason that its formidable influence needs to be factored back into the history of late Victorian Britain. This is particularly true in the case of the prince's royal visit in 1875–76 because of the imperial symbolism connected to his physical presence in India.

Picking up on the Hindu concept of a deity being incarnated in living form, British commentators described the prince as an avatar of British imperialism—identifying him as an embodiment of the otherwise amorphous entity of the British raj. William Howard Russell, for instance, noted how in Calcutta "a peculiar want of human nature has been gratified by the Prince's avatar."[74] Meanwhile, another unnamed member of the royal entourage remarked that the Prince of Wales "had shown himself to princes, soldiers, peasants and workers as the incarnation of the British Raj, which had previously been no more than a remote and abstract symbol."[75] Thanks to the vector of the virtual, this vision of the prince's imperial performance in India, with its exhilarating blend of fact and exotic fantasy, was extended and projected back to Britain as a type of "moving panorama" through the medium of illustrated journalism. It is no coincidence that Prime Minister Benjamin Disraeli seized the political initiative, on the back of the popular enthusiasm for the imperial project engendered by the media reports of the prince's royal progress, to have Queen Victoria declared Empress of India—even before the prince himself had time to set foot once more on British shores.

Notes

1. Anne Friedberg, *The Virtual Window: From Alberti to Microsoft* (Cambridge, MA: MIT Press, 2006), 9–10.
2. The prince's brother, Alfred, the Duke of Edinburgh, had made a world tour in 1870 that included an Indian itinerary. However, his visit had not warranted the same media attention, and only a small selection of images from the tour appeared in the press.
3. William Simpson, the artist working for *The Illustrated London News*, held an exhibition featuring over two hundred images called *India "Special"* at the Burlington Gallery from May 27 to July 8, 1876. Meanwhile, Sydney Prior Hall's illustrations appeared among the display *The Prince's Indian Gifts* held at the South Kensington Museum from June 22 until September 30, 1876.
4. "*The Builder*, 10 June 1876" [artist's cutting] in annotated copy of William Simpson, *India "Special"*, exh. cat., 1876, n.p., in British Library.
5. For details of the emergence of the artist-reporter as a profession, see George Eyre Todd, ed., *The Autobiography of William Simpson, R. I.* (London: T. Fisher Unwin, 1903); Mason Jackson, *The Pictorial Press: Its Origin and Purpose* (London: Hurst & Blackett, 1885); Paul Hogarth, *The Artist as Reporter* (London: Gordon Fraser Gallery Limited, 1986).
6. George Augustus Sala, *India and the Prince of Wales* [the Indian extra number of *The Illustrated London News*], London, 1875), 7. Sala had no

further involvement with the reportage of the prince's tour after the writing of this special number.

7. For a discussion of the virtuality exhibited by these other cultural forms, see Paul Arthur, *Virtual Voyages: Travel Writing and the Antipodes 1605–1837* (London: Anthem Press, 2010); Alison Byerly, *Are We There Yet? Virtual Travel and Victorian Realism* (Ann Arbor: The University of Michigan Press, 2013); Alison Griffiths, *Shivers Down Your Spine: Cinema, Museums, and the Immersive View* (New York: Columbia UP, 2008).
8. Griffiths, *Shivers* 49, 67.
9. For references to the history of imaginary travel as an artistic genre, see Mildred Archer and Ronald Lightbown, *India Observed: India as Viewed by British Artists 1760–1860*, (London: Victoria and Albert Museum, 1982), 14, 79–80, 82, 86; Arthur, *Virtual Voyages* xvii–xxii, 1–7; Byerly, *Are We There Yet?* 1–14, 25–26, 28; Griffiths, *Shivers* 40, 49, 84–86.
10. Sala, *India* 7. The trope of "armchair travel" is discussed in Archer and Lightbown, *India Observed* 14, 79–80, 82, 86; Griffiths, *Shivers* 9, 84, 86; Byerly, *Are We There Yet?* 25–26, 28, 47, 50.
11. Sala, *India* 7.
12. Declared in the concluding paragraph of the periodical's opening address, *The Illustrated London News*, May 14, 1842, 1. Cited in Byerly, *Are We There Yet?* 34.
13. For further evidence of this perceptual development, see Byerly, *Are We There Yet?* 59, 65, 79, 81, and Griffiths, *Shivers* 43–44.
14. Byerly, *Are We There Yet?* 2, 13, 23–25, 28, 59, 65–67, 70, 72, 81–82. For a related discussion of the panorama introducing a new "way of seeing" the world, see Griffiths, *Shivers* 47, 86.
15. Byerly refers to the editorial ambition outlined in the *Illustrated London News*'s inaugural address (May 14, 1842, 1): Byerly, *Are We There Yet?* 34.
16. "*Morning Post*, 29 May" [artist's cutting] in Simpson's annotated copy, *India "Special,"* 1876, n.p., British Library.
17. William Simpson, *Notes and Recollections of My Life* (handwritten memoirs in the collection of the National Library of Scotland), 1889, 296–97.
18. After his travels around India in 1859–1862, Simpson became a member of the Royal Asiatic Society, befriended the well-known authority on Indian architecture James Fergusson, and published several articles on Indian culture and theology. William Simpson, *Notes and Recollections* (1889), 234, 246; Eyre Todd, ed., *Autobiography of William Simpson*, 175.
19. Here, Jackson alludes back to the ideas presented in the first address of *The Illustrated London News*: Jackson, *Pictorial Press* 296.
20. Walter Charles Horsley, son of the more celebrated RA painter, John Callcott Horsley, was redirected to cover events in Egypt halfway through the prince's tour, leaving Herbert Johnson to provide sole coverage for the remainder of the visit.

21. Lewis Lusk, "A Famous Journalist, Sydney P. Hall, M. V. O.," *Art Journal* (London: Virtue & Co.,1905), 279.
22. See W. H. Russell, *The Prince of Wales' Tour: A Diary of India* (London: Sampson Low, Marston, Searle & Rivington, 1877).
23. In November 1876 William Simpson published a photographic album called *Shikare and Tomasha: A Souvenir of the visit of H. R. H. The Prince of Wales to India*, featuring a set of 12 photographic reproductions of his original sketches. These were chosen, with the prince's permission, from the 15 images purchased by the Prince of Wales from the artist's exhibition.
24. Studies investigating the role of art and photography in constituting an "imagined India" include Archer and Lightbown, *India Observed*; Pratapaditya Pal and Vidya Dehiejia, *From Merchants to Emperors: British Artists and India 1757–1930* (Ithaca, NY: Cornell UP, 1986); James Ryan, *Picturing Empire: Photography and the Visualization of the British Empire* (Chicago: University of Chicago Press, 1997); Romita Ray, "The Memsahib's Brush," in *Orientalism Transposed: The Impact of the Colonies on British Culture*, eds. Julie F. Codell and Dianne Sachko Macleod (Aldershot, UK: Ashgate, 1998); Giles Tillotson, *The Artificial Empire: The Indian Landscapes of William Hodges* (Richmond, UK: Curzon, 2000); Hermione de Almeida and George H. Gilpin, *Indian Renaissance: British Romantic Art and the Prospect of India* (Aldershot, UK: Ashgate, 2006).
25. Sala, *India* 7. One of the founding principles of *The Illustrated London News* had been that the "the progress of illustrative art" now allowed "the geography of mind" to be "mapped out . . . with clearer boundaries and more distinct and familiar intelligence that it ever bore alone": "Our Address," *The Illustrated London News*, May 14, 1842, 1.
26. Archer and Lightbown, *India Observed* 11, 87.
27. Griffiths describes how the composite panoramas of the 1840s similarly blended standard topographical scenes with "metonymically rendered narrative peoples": Griffiths, *Shivers* 51–55.
28. Griffiths, *Shivers* 69–71.
29. Simpson refers to this artistic continuity in his memoirs, telling how he deliberately prepared for his initial visit by "looking over books about India, such as Daniel's [sic] to see what had been done already—and to get hints as to places I ought to visit": Simpson, *Notes and Recollections* 109.
30. For details of their involvement in the topographical art tradition, see Wilcox, *Edward Lear*.
31. Simpson's precise dating of his sketch is not in itself unusual, as this was standard practice for topographical artists, such as Edward Lear, who sometimes even recorded the time of day a drawing was made. The factors that quintessentially distinguish the work of the artist-reporter are the inclusion of topical references and the inherent news value of the chosen image.

32. Byerly similarly notes a developing focus "on the mechanisms of travel" by the end of the nineteenth century: Byerly, *Are We There Yet?* 25.
33. In this case, they were even a matter of controversy at the time. Tacitly acknowledging the potential for outrage, Russell records that the Brahmins let the local Hindu community believe that the prince went to Elephanta to "worship the Deity": Russell, *Prince of Wales' Tour* 165.
34. Byerly, *Are We There Yet?* 67; Griffiths, *Shivers* 70.
35. For an insightful, postcolonial reading of the content of the illustrated press coverage of the prince's Indian visit, see H. Hazel Hahn, "Indian Princes, Dancing Girls and Tigers: The Prince of Wales's Tour of India and Ceylon, 1875–1876," *Postcolonial Studies* 2, no. 2 (2009): 173–192.
36. "The Builder, 10 June 1876" [artist's cutting], Simpson, *India "Special"*, n.p.
37. Anon., "Our Sketches from India," *The Illustrated London News*, February 12, 1876, 162.
38. Sala, *India* 7.
39. For in-depth analysis of this phenomenon, see Byerly, *Are We There Yet?* 1, 4–5, 14–15, 17, and Griffiths, *Shivers* 9, 37–43.
40. Julie Codell, in her pioneering investigations of the workings of the nineteenth-century press, notes how the periodicals effectively produced a "virtual printed space" shared between Britain and its colonies but does not explore the idea further, Julie F. Codell, ed., *Imperial Co-Histories—National Identities and the British Colonial Press* (Madison, UK: Fairleigh Dickinson UP, 2003), 15, 18.
41. Sala, *India* 5.
42. These shared virtual attributes are identified by Byerly, *Are We There Yet?* 1–3, 5, 15, 18–19, 23, 30, 39–41, 55–6, and Griffiths, *Shivers* 1–5, 37, 40, 43, 49, 69, 73, 83.
43. Byerly notes how "an illusion of physical movement" helps "the reader or viewer to feel fully engaged in the fictive environment," while also being an "essential" factor in "creating that sense of locatedness": Byerly, *Are We There Yet?* 14, 16.
44. For a discussion of the critical role of "mediated presence" in creating virtuality, see Byerly, *Are We There Yet?* 5, 15–16, and Griffiths, *Shivers* 3–4, 40, 83.
45. Michael Fried, *Menzel's Realism: Art and Embodiment in Nineteenth-Century Berlin* (New Haven, CT: Yale UP, 2002), 13, 19.
46. Lusk made direct comparisons between Hall's and Menzel's oeuvres, noting they shared an aptitude "for free vital pen-line": Lusk, "A Famous Journalist" 278–80.
47. Fried himself tentatively advances this claim, observing at one point, "I stop just short of calling Menzel's drawings a virtual portrait of [Mr.] Puhlmann, though that would be one possible implication of my remarks." He considers the virtual implications of writings on the

"aesthetics of empathy" associated with Menzel's artistic effects. Fried, *Menzel's Realism* 3, 36.

48. "*The Builder*, 10 June 1876" [artist's cutting], *India "Special"*, n.p., British Library.
49. Fried theorizes that for Menzel sight is always conditioned by "the intuition of previous experiences, non-visual sensory modalities and aspects of the world, manual and other bodily operations" and discusses how he communicates these latent influences to the viewer: Fried, *Menzel's Realism* 2–3, 13, 24, 28, 31–32, 34, 36, 50–51.
50. Fried, *Menzel's Realism* 13–14, 19, 20, 22, 28, 31–32, 34, 39, 42, 50.
51. Fried, *Menzel's Realism* 19–20, 22.
52. Baudelaire used Guys as the exemplar of a new class of artist he identified as "the Painter of Modern Life": Charles Baudelaire, *The Painter of Modern Life and Other Essays*, trans. and ed. John Mayne (London: Phaidon Press, 1964), 16.
53. Byerly reaches the same conclusion, identifying a rhetorical strategy whereby realist authors, such as George Eliot, use the figure of the "hypothetical traveler" as a substitute for the reader "in much the same way an avatar represents the player of a computer game." She also notes that this trope offers a variant on the theme of the "flaneur": Byerly, *Are We There Yet?* 1–3, 10, 27–28, 72–73, 81.
54. Leonard Bell, "To See or Not to See: Conflicting Eyes in the Travel Art of Augustus Earle," in *Orientalism Transposed*, 121.
55. The special production of an Anglo-Indian periodical called the *Royal Tourist*, featuring the illustrations from *The Graphic* and *The Illustrated London News*, distributed in India contemporaneously with the prince's visit, represents a conscious extension of this effect. In her historical study, Hahn explores the significance placed on his tourist identity and the way he was both "at the centre of attention and a spectator": Hahn, "Indian Princes" 175–81.
56. See *The Graphic*, January 8, 1876, 29.
57. This idea is central to Fried's theories concerning the "art of embodiment." In one example of Menzel's work, he refers to the "phantomlike evocation of a viewer—in the first instance the artist": Fried, *Menzel's Realism* 13–14, 26, 34, 37–39, 50.
58. Simpson, *Notes and Recollections* 296–97.
59. Griffiths, *Shivers* 73–74.
60. For the workings of this progression and the "mimetic fallacy," see Wilcox, *Edward Lear* 35–37, and Nancy Armstrong, "Realism before and after Photography: The Fantastical Form of a Relation among Things," in *A Concise Companion to Realism*, ed. Matthew Beaumont (Oxford: Wiley-Blackwell, 2010), 102–9.
61. Sontag's line is quoted in James Ryan, *Picturing Empire: Photography and the Visualization of the British Empire* (Chicago: University of Chicago Press, 1997), 72. For the prevalence of this approach, see Nancy Armstrong, "Realism before and after Photography" 107–16; James Ryan,

"Images and Impressions: Printing, Reproduction and Photography," in *The Victorian Vision: Inventing New Britain*, ed. John M. Mackenzie (London: V&A Publications, 2001), 223, 227, 234; Peter D. Osborne, *Travelling Light: Photography, Travel and Visual Culture* (Manchester: Manchester UP, 2000), 3, 9, 19, 22; Wilcox, *Edward Lear* 36–7; Ryan, *Picturing Empire* 16, 26, 45–47, 72, 214; Jennifer Green-Lewis, *Framing the Victorian: Photography and the Culture of Realism* (Ithaca, NY: Cornell University Press, 1996), 6, 25, 100, 106, 110; Pal and Dehiejia, *From Merchants* 206.

62. Anon., "Photography in India," *British Journal of Photography*, October 8, 1875, 492.
63. A view of Jaipur by Deen Dayal, *The Chandra Mahal of the City Palace*, dated to 1876, is illustrated in Vibhuti Sachdev and Giles Tillotson, *Building Jaipur: The Making of an Indian City* (London: Reaktion Books, 2002).
64. The extra large volumes produced for the Prince of Wales remain in the Royal Collection. Copies of the smaller commemorative publication, *Bourne and Shepherd's Royal Photographic Album of Scenes and Personages Connected with the Progress of H. R. H. The Prince of Wales through Bengal, the North West Provinces, the Punjab and Nepal* (Calcutta: Bourne & Shepherd, 1876), reside in the British Library and the National Library of India.
65. This was the company's standard business practice. For details see Gary D. Sampson, "Unmasking the Colonial Picturesque: Samuel Bourne's Photographs of Barrackpore Park," in *Colonialist Photography: Imag[in]ing Race and Place*, eds. Eleanor M. Hight and Gary D. Sampson (London: Routledge, 2002), 87, 102–3.
66. See Osborne, *Travelling Light* 3, 22; Ryan, *Picturing Empire* 16, 26, 214; Ryan, "Images and Impressions" 227, 234.
67. Heinz K. Henisch and Bridget A. Henisch, *The Photographic Experience 1839–1914: Image and Attitudes* (University Park: The Pennsylvania State UP, 1994), 13.
68. Archibald Forbes, "*Daily News*, 29 May 1876," [artist's cutting] in Simpson's annotated copy, *India "Special"*, 1876, n.p., British Library.
69. Harry V. Barnett, "The Special Artist," *The Magazine of Art* (London: Cassell and Company, 1883), 166.
70. Charles Baudelaire, *Painter of Modern Life* 16.
71. E. R. Jaensch, *Eidetic Imagery and Typological Methods of Investigation* (Westport, CT: Greenwood Press, 1930, reprinted 1970), 1–5.
72. For the nineteenth-century view on the importance of the reportorial exercise, see Simpson, *Notes and Recollections* 296–97.
73. Griffiths, *Shivers* 74. Citing Walter Benjamin's arcade theories, Byerly describes the panorama as an example of phenomena that "create a transitional space between past and present, distance and interior, nature and art": Byerly, *Are We There Yet?* 32–33. Comparing fictional and nonfictional travel accounts, Arthur notes "the extent to which observation

was influenced by received traditions and fantasy" and that "the blurring of fact and fiction was not unique to imaginary voyages": Arthur, *Virtual Voyages* 3–4. These factors are undoubtedly at work in the pictorial reportage of the prince's Indian journey.

74. Russell, *Prince of Wales' Tour* 352.
75. Quoted in Geoffrey Bennett, *Charlie B: A Biography of Admiral Lord Beresford of Metemmeh and Carraghmore* (London: Peter Dawnay, 1968), 56.

Chapter 10

Steampunk Technologies of Gender: Deryn Sharp's Nonbinary Gender Identity in Scott Westerfeld's *Leviathan* Series

Lisa Hager

N.B. The images associated with this chapter are housed in the digital annex at www.virtualvictorians.org.

As a discipline, Victorian studies has a certain fondness for pointing out the myriad connections between the nineteenth century and our own time. We do so not only to make the vital case for the relevance of our work to current cultural concerns, but also to bring into relief how twentieth- and twenty-first-century cultures have understood themselves in relation to the Victorian. As literary genres overtly concerned with understanding and reimagining nineteenth-century literary culture, neo-Victorian fiction and, in particular, steampunk fiction are central nodes for this sort of cultural and critical work. First named by K. W. Jeter in 1987,[1] steampunk, a subgenre of science fiction and fantasy in which alternate histories of the nineteenth century dramatically reshape the past, present, and future, has become increasingly popular in mainstream culture as both a literary genre and a fan subculture, to the point that one can now purchase mass-produced steampunk costumes from most American party shops. As an alternate history genre, steampunk literature builds its worlds by answering the "what if?" question of science fiction and fantasy by tweaking nineteenth-century history and literature and engaging with cultural discourses of the day.

Steampunk often revises and questions Victorian gender norms, even as it acknowledges the power of those norms. In steampunk fiction, this reconfiguration of nineteenth-century gender roles is deeply enmeshed in the genre's focus on retrofuturistic technology from which it earns the "steam" part of its name. Moreover, steampunk fiction's play with and "punking" of the connections between gender and technology can also reveal the work required to maintain the Victorian gender binary, as well as our own. Bringing this subversive steampunk approach to gender to young-adult fiction, Scott Westerfeld's *Leviathan* trilogy follows the exploits of Deryn Sharpe as she plays a key role in the war between Darwinist Western Europe, whose technology relies on genetic modification of living creatures, and Clanker Germany and Austria-Hungary, whose technology relies on gears and steam engines. Deryn becomes a boy named Dylan Sharpe in order to become a midshipman aboard the *Leviathan*, a genetically engineered whale zeppelin that is part of England's Air Service. In the process, she aids and develops feelings for young Prince Aleksandar of Hohenberg, who is trying to stop the world war sparked by the death of his parents.

Throughout *Leviathan* (2009), *Behemoth* (2010), and *Goliath* (2011), Westerfeld uses the gender transgressions of both Deryn and Alek—and the technology that makes those transgressions possible—to rewrite prevailing narratives of gender and romance in young-adult fiction and to interpret Victorian ideologies of gender in terms of their instability. In the trilogy, Westerfeld combines the Victorian boys' adventure novel with the courtship novel. By putting these two genres in conversation with each other in the adventures of Deryn Sharpe, the author invites readers to reconsider the extent to which we continue to read novels with women and girl protagonists chiefly for the courtship plot, popularized in the triple-decker novels of the nineteenth century and still a pervasive force in contemporary young-adult literature. This, in turn, suggests a more complex picture of Victorian culture that is attuned to the diverse ways in which nineteenth-century gender identities function in relation to separate-spheres ideology.

Westerfeld grounds Deryn's gender performance in steampunk's visual performance of the Victorian. One of the primary (and perhaps the most obvious) methods by which steampunk enacts and revises the Victorian is through its deployment of period aesthetics and its thematization of the relationship between aesthetics and technology. Given the genre's diversity of setting, from Victorian London to nineteenth-century Istanbul, this visual style plays a key role in establishing the genre as such. Mike Perschon argues that this quality becomes even more important in understanding the genre in the popular imagination, describing "steampunk as an array of visual markers which, when combined, constitute the look popularly understood as steampunk."[2] In commenting on recent steampunk literary work, Ann and Jeff VanderMeer argue that "steampunk has indeed become an aesthetic toolbox for a range of approaches."[3] As these comments suggest, aesthetics are perhaps the most recognizable feature of steampunk as literary genre and subculture. When trying to explain steampunk to students, colleagues, or friends, I find that most are at least slightly familiar with some of the popular visual markers of steampunk—corsets and cogs. While these aesthetic signs have gotten so ubiquitous as to give rise to parodies, as exemplified by the website Regretsy's "Not Remotely Steampunk" feature and the popular YouTube video and song "Just Glue Some Gears on It (and Call It Steampunk)" by Reginald Pikedevant, their very pervasiveness within steampunk literary and material culture indicates their importance. Though not present in all steampunk texts or objects, they do metonymically represent both steampunk's reading of nineteenth-century culture and its central concern with the interplay between aesthetics, gender, and technology.

Steampunk's juxtaposition of these visual markers reveals the connections between them. Corsets, with their simultaneous containment of the female form when tightly laced and support of women's bodies when more loosely worn, represent Victorian England's anxiety over the Woman Question. Cogs, on the other hand, are closely associated with the Industrial Revolution, gentlemen inventors, and, to borrow Thomas Carlyle's term, captains of industry. These two common symbols of steampunk aesthetics seem to represent a clear binary that opposes fashion and technology, women and men, bodies and minds. However, much of steampunk literature rejects this opposition's inherent hierarchy by valuing both equally. As suggested by objects like noted steampunk maker Jake von Slatt's modifications that reenvision the computer as being powered by gears and steam, as well as by the popularity of wearing one's corset over one's clothes in steampunk,[4] this technology—be it digital or sartorial—eschews smooth uniform modernity in favor of an approach that celebrates the bespoke, the gritty, and the unfinished. This ethos displays the inner workings of technology rather than hiding them behind glossy facades like so many of our modern machines, as well as the interpretive work involved in our constructions of the nineteenth century.

In Westerfeld's *Leviathan*, Deryn and Alek embody this steampunk style both in their use of fashion and biomechanical inventions and in the assemblages of their unconventional gender identities, recalling Lev Grossman and Gary Moskowitz's description of steampunk: "Steampunk isn't mass-produced; it's bespoke and unique, and if you don't like it, you can tinker with it till you do."[5] By tinkering with their performances of gender, these two characters evoke a steampunk version of Donna Haraway's cyborg, which she defines in terms of its hybridity: "A cyborg is a cybernetic organism, a hybrid of machine and organism, a creature of social reality as well as a creature of fiction."[6] The cyborg occupies the intersection of several binaries and, in so doing, reveals the permeability of the boundaries between such categories—animal and machine, feminine and masculine, fiction and reality. This cyborg hybridity is at the heart of steampunk's understanding of technology. As steampunk writer Professor Calamity remarks in "My Machine, My Comrade,"

> Steampunk seeks to find a relationship with the world of gears, steel, and steam that allows machines to not only co-inhabit our world but to be partners in our journey. To be born, age, and die like we all must, that is not only true of humans, plants, rivers, animals but also of machines. This may be a crucial realignment of our relationship to the world, man-made and natural.[7]

In Westerfeld's trilogy, this hybrid relationship with technology takes the form of a consistent attention to and redefinition of the identities of "boy" and "girl." In particular, it shows how the related identities of "soldier" and "airman" intersect and ultimately dissolve the rigid boundaries between genders through the mobilization of steampunk technologies. In short, Deryn and Alek offer us a reading of contemporary gender that complicates our own readings of Victorian gender, standing in for us as readers who view Victorian culture through steampunk's reimagined re-creation of the past.

Moreover, just as steampunk design aesthetic centers around making visible the inner workings of technology, steampunk literature is conscious and critical not only of how it seeks to inhabit the Victorian but also of the ways in which its twenty-first-century point of view informs that world. In doing so, steampunk draws attention to the fact that any recreation, including scholarly readings, of the Victorian era ultimately rests on a virtual reality of the period that is shaped by a constantly shifting relationship among readers, creators, and texts. In Westerfeld's trilogy, this self-conscious world-building of an alternate virtual Victorian England reveals itself most clearly in Deryn Sharpe's purposeful reconfiguration of gender norms.

In depicting Deryn's cross-dressing, Westerfeld reads Victorian gender codes in terms of clothing and unsettles the supposed stability of such codes by highlighting their malleability. Deryn begins the novel rejecting the limitations of feminine domesticity as represented by confining clothing and domestic space: "Her mother and aunties were waiting there . . . ready to stuff Deryn back into skirts and corsets. No more dreams of flying, no more studying, no more *swearing*!"[8] Deryn's dislike of feminine clothing is explicitly tied to the activities it prohibits and the gendered identity it requires; clothes signify identity, which in turn determines one's social role. In the "skirts and corsets" of her[9] women relatives, Deryn would be forced to conform to a gender identity that does not match her authentic sense of self.

In highlighting the highly gendered nature of nineteenth-century clothing, Westerfeld is also able to demonstrate how that seemingly conservative code enables Deryn to take on a masculine identity once she changes her clothes—proving the cliché that clothes do make the "man" (or rather, airman). Once aboard the *Leviathan* in her uniform, Deryn revels in her new clothes and gender identity:

> Her airman's uniform was miles better than any girls' clothes. The boots clomped gloriously as she stormed to signals practice or firefighting drills, and the jacket had a dozen pockets, including special compartments for her command whistle and rigging knife. And Deryn didn't mind the constant practice in useful skills like knife throwing, swearing, and not showing pain when punched.[10]

As this passage demonstrates, the airman's uniform enables Deryn to be more fully herself. It allows her to perform the duties of her position and to inhabit the active role of a young man.

The shift between the dysphoria Deryn experiences while in women's clothes and the gender-affirming pleasure she experiences while in men's clothes marks her as a transgender character.[11] Like many transgender people, Deryn uses clothing to communicate her gender to others via coded aesthetic signs—and to own her identity more securely. Like the specialized gender-affirming prosthetics used by some transgender people to confirm their internal sense of self, Deryn's airman uniform is not a disguise worn to hide her real identity, but rather steampunk fashion technology, with its tools specifically geared toward her work aboard the *Leviathan*, enables her to become who she truly is.

Reflecting steampunk's push to occupy the boundaries between binaries, Deryn's gender cannot be easily classified as feminine or masculine; instead, she occupies what we might call a *genderqueer* or *gender-fluid* space in which neither binary label can suffice. This transgressive gender identity makes visible the workings of gender itself in much the same way as a cog-displaying computer plays with the boundary between the analog and the digital, as J. Jack Halberstam argues in *Female Masculinity*: "Female masculinity actually affords us a glimpse of how masculinity is constructed as masculinity. In other words, female masculinities are framed as the rejected scraps of dominant masculinity in order that male masculinity may appear to be the real thing."[12] In highlighting the breakdown of the man/woman binary, Deryn's genderqueer identity insists that we reevaluate our understanding of Victorian gender. We must question whether ideologies of gender can encompass the complexity of any sort of humanity, be it virtual or real, and must therefore question our continued adherence to this dysfunctional binary.

Furthermore, Deryn's airman uniform makes her part of the ship and ties her to the steampunk technological marvel that is the *Leviathan*. A creation of England's Darwinist boffins (that is, scientists), the *Leviathan* is a genetically modified hydrogen-breathing whale in the British Air Service that flies and does battle using both engines and the various other modified species that make up its fully functioning ecosystem, what Deryn calls "a whole tangle of beasties," including fléchette bats that drop metal spikes and bees whose nectar feeds the bacteria that produce hydrogen.[13] A thoroughly interrelated ecosystem of machines and animals, the airship becomes even more so after Alek and his crew customize their Austrian military mechanical walker's engines to fit the *Leviathan* after its own engines are irreparably damaged in battle. This airship's combination of the mechanical and the biological, the inanimate and the animate, exemplifies steampunk technology's interrogation of the boundary between them.[14]

The ease with which Deryn quickly becomes an essential part of this hybrid ecosystem is bound up in the ways in which this airship enables her to be her authentic self in terms of gender: "The *Leviathan* felt like part of Deryn now . . . the first place where no one had ever seen her in a skirt, or expected her to mince and curtsy."[15] Deryn's self-assured comfort on this airship comes from the alignment of her interior sense of identity and the way in which her fellow crewmen see her. This alleviates the gender dysphoria she experienced when dressed as a girl. Indeed, Deryn's belongingness to the *Leviathan*'s human and animal ecosystem becomes literally true when she injures her knee while gliding on a mission and is given a "half plant and half animal" compress to heal her torn ligaments.[16] Much like the uniform she wears, this "wee fabricated beastie"[17] serves to link her to the ship's interdependent community of animals and humans. This steampunk technology tinkers with the nature of life itself, reflecting the way in which Deryn's transgressive identity questions the oppositional gender binary that structures so much of steampunk Victorian society.

Throughout the books, Deryn becomes the lynchpin of the *Leviathan*'s human ecosystem, largely through her quick and unorthodox critical thinking. Once aboard the ship, she distinguishes herself in her knowledge of aeronautics, skill in handling the various animals that make up the *Leviathan*'s ecosystem, and bravery in risking her life for her missions and in protecting her crewmates and friends—so much so that she earns the Air Gallantry Cross, one of the Air Service's highest honors, for saving a fellow crewman.[18] Deryn's pivotal role in the *Leviathan*'s crew and its missions is perhaps best demonstrated when the astute journalist Miss Adela Rogers nicknames her the *Leviathan*'s "bell captain" because

> the most important person in a hotel isn't the owner, or the manager, or even the house detective. It's the *bell captain*. He's the one who knows where the bodies are buried . . . And everyone glances at you when they've got a pickle to deal with. Dr. Barlow, Prince Aleksandar, even that crusty old count—they all want to know what the bell captain thinks.[19]

As Rogers implies, Deryn, though a lowly midshipman, plays a pivotal role in dealing with many difficult and delicate political and military situations the *Leviathan* encounters as Europe and the world are pulled into a world war. This dynamic creates creating an alternate nineteenth-century European history that depends upon binary-eroding technology and gender.

Critically, it is Deryn's movement between the masculine and the feminine that enables her to take on this vital role, specifically regarding her affection for Prince Aleksandar. As with Deryn's claiming of a masculine identity aboard the *Leviathan*, Westerfeld explains her first attraction to Alek in terms of technology: "Where Alek's arms had wrapped around her shoulders an odd kind of tingling was left behind—like the crackle along the airship's skin when distant lightning kindled the sky."[20] Deryn conceptualizes the effect that Alek's touch has on her in terms of the *Leviathan*'s response to electricity, using her first love, the ship, to understand her second. As Deryn comes to understand that she loves Alek, even though his claim to the throne of the Austrian-Hungarian Empire means they can never be together, Westerfeld codes her love for Alek as feminine, conflating the class and gender transgressions of her affection: "Deryn realized that she'd let her voice go all squeaky. That was what came of thinking too hard about Alek—it turned her pure dead girly sometimes."[21] Though Westerfeld here plays on conventional notions of gender that ascribe weakness and sentiment to femininity, which Deryn's higher voice indicates, the author does not make his character's feminine qualities a source of weakness. Nor do they conflict with her masculinity as a midshipman. Instead, the concern that Deryn expresses here leads her to seek out Alek in Istanbul, where he is helping revolutionaries overthrow the sultan, and it is Deryn's idea of using spice bombs that is key to the revolution's success. In this escapade, Deryn's affection for Alek is the catalyst for her heroic actions that ultimately slow the spreading war by keeping the Ottoman Empire neutral. Her feminine side is thus a source of strength and is essential to her "bell captain" role.

Given Deryn's commitment to the *Leviathan* and the life it has given her, it is the loss of that life and the censure of the crew that she most fears when details about her past are discovered by a news reporter. She does worry about her and Alek's relationship when her secret is in danger of being revealed, but she is equally concerned about the effects of that revelation on her shipmates and on the military branch they serve. To avoid such scandal for the Air Service, Deryn accepts a lady scientist's offer to join the Zoological Society, which functions as a secret diplomatic organization working for the good of the British Empire: "If she were found out, it would humiliate her officers and shipmates . . . In the Society, she reckoned, having more than one identity wouldn't be a problem at all. Dr. Barlow had even joked that Deryn might need to disguise herself as a girl, every now and then."[22] Though Deryn knows that this decision means that she loses both her "home" and Alek, she chooses a life that will enable her to avoid tarnishing the reputation of those whom she respects—and a life that will give her meaningful work. Importantly, her new position as Dr. Barlow's personal assistant gives Deryn the freedom to have multiple identities and move between feminine and masculine dress and roles. Indeed, it is this very movement that makes her a valuable assistant, as it demonstrates her ability to keep secrets and inhabit a variety of identities.

Deryn's relationship with Alek evolves to reflect this multiplicity of identities as the prince reevaluates his gender and class prejudices in order to understand Deryn's significance in his life. He begins the novel as a sheltered royal child who is uncomfortable around commoners and unprepared for the world outside the walls of his parents' estate. Before Alek knows about Deryn's complicated gender identity, he sees her as a model for what he might have been without the trappings of royalty: "'*There's* the boy I want to be—or would, if I hadn't been born such a hopeless prince.'"[23] For Alek, Deryn's bravery and skill in her role as an airman is to be admired for fully exemplifying ideal masculinity—much more so than his own limited knowledge as a prince. Alek's assessment of Deryn's boyhood suggests a version of gender that relies on a person's performance rather than society's interpretation of his or her biological characteristics. Moreover, by having Alek admire Deryn's derring-do and quick thinking as traits of the ideal young man, Westerfeld creates a boys' adventure series in which one of the central masculine characters challenges essentialist notions of gender identity and instead argues for a self-determining model of gender.

Once he learns Deryn's secret and develops a strong attachment to her, Alek must not only shift his understanding of gender but also ultimately reject his elevated class status. Alek's first reaction upon figuring out that Deryn had previously been forced into the identity of a girl reveals the power that transphobia had over Victorian ideologies of gender—as well as current conceptions. After putting the pieces together, Alek obsessively repeats to himself that Deryn is not what she seems: "She was a girl. Her name was Deryn Sharp, and she was a girl disguised as a boy."[24] He even goes so far as to tell her that "You're not even a real soldier" and chides her for the tears that his words bring to her eyes.[25] Here, Alek insists that Deryn's assigned-at-birth gender precludes her from having any access to the identity of soldier or boy, and that her weak and feminized display of emotion only serves to confirm that she must be a girl. In this biological essentialism, Alek positions "girl" as inferior to "boy," illustrating the extent to which Victorian separate-spheres ideologies, which sought to cast the public world as men's domain and the private domestic space as women's realm, ultimately rest upon thinly disguised misogyny.

In Deryn's case, because her identity mixes masculinity and femininity, this prejudice also reflects a fear of inauthentic and deceptive gender in the form of transmisogyny, which Julia Serano defines in regard to transgender and transsexual women in *Whipping Girl: A Transsexual Woman on Sexism and the Scapegoating of Femininity*: "When a trans person is ridiculed or dismissed not merely for failing to live up to gender norms, but for their expressions of femaleness or femininity."[26] Though Serano uses this term in order to get at the double-bind oppression that transwomen face, it is also applicable to gender identities that occupy the space between masculinity and femininity since these people can be subject to similar oppression for their femininity. In this moment, Alek cannot wrap his mind around the idea that the boundaries between boyhood and girlhood are permeable, and his insistence that Deryn is a girl and therefore cannot be a soldier represents an attempt to reinforce those boundaries, despite the authentic experience of his most trusted friend.

Yet Deryn refuses to simplify her complex identity. It is Alek who changes to accept her as she is, by giving up his birthright and reconceptualizing gender. Having to reconcile the seemingly incompatible identities of girl and soldier, Alek's appreciation for his friend's loyalty and bravery eventually enables him to see that she is "a real soldier—quite a good one, in fact."[27] His shifting mindset is largely due to an epiphany that he experiences on viewing a silent film (*The Perils of Pauline*) in which the heroine rescues herself from a runaway hot-air balloon, echoing Deryn's own quick thinking at the beginning of the book. This media representation of an active woman with the agency to take care of herself in perilous situations forces Alek to realize that Deryn can be both a girl with romantic feelings toward him and a daring airman. As he tells Deryn, "The point is that it's [the film] terribly popular. So even if American women aren't piloting balloons yet, at least they want to."[28] *The Perils of Pauline* has this effect on Alek because it allows him to visually experience a world that does not yet exist and to imagine how his friend could honestly be herself in that world. It is the immersive qualities of film, which in this steampunk world are shot using mechanical walkers that have cameras mounted on them, that make redefined gender roles seem possible. In an alternate world within the alternate world of the novel, the film gives Alek a virtual experience of a more expansive and flexible notion of gender. In linking Alek's new understanding of Deryn to this new form of steampunk media, Westerfeld not only suggests the importance of information technology to our understanding of gender identity but also highlights the importance of nineteenth-century media technology to gender and sexual minorities of the day—ideas exemplified in our own time by the widespread use of social media like Tumblr and YouTube for activism in the LGBTQIA community.[29]

This focus on media technology's relationship to gender diversity is further underscored by Alek's decision to relinquish his royal ambitions. Alek realizes that he values his friendship and burgeoning romance with Deryn more than his contested claim to the throne, since it is together that they accomplish the most good for themselves and the world: "'Maybe I wasn't meant to end the war, but I was meant to find you. I know that. You've saved me from not having any reason to keep going.'"[30] After telling Deryn that he is committed to their partnership, he destroys the pope's scroll, the ultimate example of archaic information technology, that would legitimize him as heir to the throne. In its ability to make Alek the ruler of millions with a few words from one of history's most powerful authority figures, the scroll represents a top-down system of power and information, which steampunk's crowdsourced ethos refuses. This act symbolizes not only Alek's acceptance of Deryn but also his movement away from the rigid class distinctions and gender roles that are increasingly outmoded in this hybrid steampunk universe. In addition, Alek's commitment to his friend illustrates steampunk-specific reading practices as he reinterprets his own story; he exemplifies the modern sensibilities that we, as readers, bring to the table when we consider the nineteenth century.

The "Bonus Chapter" to *Goliath* ties up these thematic threads and firmly establishes Deryn and Alek's relationship as one in which gender identities will constantly be in the state of flux that both of them prefer. This short vignette finds our two heroes, along with Dr. Barlow and Count Volger, at a New Year's Eve masquerade ball put on by the London Zoological Society. Exemplifying Bakhtinian carnival, this fancy-dress ball represents a moment of frivolity and topsy-turviness, as Count Volger, who has dressed as a Darwinist message lizard—the genetically modified lizards that act as living voice recorders and messengers—though he detests the creatures, tells Alek: "One must always be ready to mock . . . Otherwise, politics becomes unbearable."[31] Such a ball, welcoming critique of and play with social conventions, provides the perfect setting for Westerfeld's final scene, since this is one of the few moments in which Deryn and Alek's shifting identities can be made public.

Ending on a note of gender destabilization, Westerfeld has both Deryn and Alek dress as women as the result of a wager. The wager itself foreshadows this unsettling of gender binaries. As Alek notes, "One moment, he and Deryn had been having a perfectly reasonable discussion on the merits of the two sexes—strength, endurance, tolerance of pain—and then suddenly he had said something unforgivable and Deryn was challenging him to an arm-wrestling contest."[32] He clings to some of his old notions about the division of the sexes, only for Deryn's physical strength to prove him unequivocally incorrect. As the loser, Alek must dress in women's clothing for the ball and ends up with a rather old-fashioned dress, complete with bustle. When asked who he is, Alek quickly decides that he is Ada Lovelace, "one of the great Clanker [British] boffins of the last century."[33] In claiming the identity of Lovelace, Alek highlights his own contradictory position as a former Clanker prince now working for the British organization that fabricates so many Darwinist creatures. He also ties in his own subversive political position to Victorian computing technology; Lovelace was the first person to fully understand the implications of Charles Babbage's difference engine and a mathematician whose work, in real life, was clearly limited by the slim opportunities for intellectual women of her day.

While playing the part of Ada Lovelace, Alek more fully understands the work of gender performance of which Deryn must be constantly aware. In this costume with its bustle, heels, and parasol, Alek considers how fully Deryn enacts masculinity:

> For a moment, he wondered at all the adjustments, small and large, that Deryn must have made in order to carry off her deception. The way she walked, talked, and stood, along with every social nuance, all of it had to be considered every second of every day. It was incredible to have succeeded at something so difficult.[34]

Alek not only appreciates the lengths to which Deryn has gone in order to honor her true identity, but also admires and even loves her for it: "Not that he minded seeing her in a jacket and trousers every day. It was part of the frisson of their romance."[35] Recalling Deryn's electric response to Alek earlier in the book, Alek's comments demonstrate that Deryn's gender fluidity is no longer any sort of impediment to their relationship, but instead has become part of the attraction between them, the "frisson of their romance."[36] Having experienced just a small bit of what Deryn's gendered life is like, Alek is made fully aware of how he now finds erotic pleasure in her nonbinary subversion, and he values her so highly that he thinks her "entirely worth throwing away an empire for."[37] Deryn's genderqueer identity is no longer coded as strange or aberrant behavior but simply part of who she is.

Further marking the mutual respect that characterizes their relationship, Alek, in contrasting his costume to Deryn's, is able to see the seemingly contradictory parts of her identity as a coherent whole. Deryn uses the opportunity of the masquerade to dress as a modern woman:

> Deryn was in the sort of evening dress that fashionable young women-about-town wore . . . Alek looked down at his own dress, so formal and old-fashioned with its fussy bows and bustle. He suddenly felt frumpy, whereas Deryn was positively stylish. Her short hair and slim figure, the core of her disguise as a midshipman, no longer looked masculine at all.[38]

In this play between pretense and reality (as Deryn's attire both is and is not a costume), Deryn's genderqueer identity is again tied to fashion—this time in the form of a ladies' gown, complete with fabricated peacock feather indicating her Darwinist allegiances. In the illustration of this scene, though Deryn appears feminine, Alek himself looks almost as womanly. In addition, it is clear from the postures of the two characters that Deryn has pulled Alek in for a kiss, much to his surprise. Since both characters are in feminine dress, the drawing allows us to see two women kissing but also to see two men kissing while dressed in women's clothes. In this final scene, then, Westerfeld mobilizes the Victorian fancy-dress ball to blur and play with gender and to suggest the fluidity within the supposedly firm boundaries between nineteenth-century masculinity and femininity.

For Deryn, the feminine skirts that she so disliked at the beginning of the novel no longer connote only women's constrained domestic role; instead, like Deryn's masculine dress, they are disguises to be deployed strategically. She even admits, "It's not as bad as I remember, being stuffed into a dress."[39] This resolution of Deryn's adventure plot figuratively brings together the corset and the cog, as Deryn can now use both at will. She can enjoy dressing as a young woman, knowing that she'll be able to wear pants the next day to perform her duties for the Zoological Society. In this way, Westerfeld revises Deryn's own initial reading of femininity to accommodate the ways in which women, like the biological engineer Dr. Barlow, command respect and demonstrate keen intellects. This prompts modern-day readers to incorporate unconventional gender expressions into their conceptualization of Victorian women.

Deryn Sharpe's genderqueer identity and the steampunk fashion and technology she uses to inhabit that identity highlight the subversive elements of Victorian dogma. Highly gendered roles and clothing are the very tools with which Deryn subverts strict gender binaries. Thus the *Leviathan* series suggests a vision of Victorian gender politics that identifies transgressive elements enmeshed within the most conservative ideologies, challenging modern readers' ideas of nineteenth-century repression and encouraging readers to question any clear-cut opposition between masculinity and femininity. In making such challenges, the series also argues for the centrality of steampunk to Victorian studies in its ability to make visible the work of scholars and steampunks alike in constructing the period.

Notes

1. K. W. Jeter, "Letter," *Locus* 20, no. 4 (1987): 57.
2. Mike Perschon, "Steam Wars," *Neo-Victorian Studies* 3, no. 1 (2010): 128.
3. Ann VanderMeer and Jeff VanderMeer, "What Is Steampunk?" in *Steampunk II: Steampunk Reloaded*, ed. Ann VanderMeer and Jeff VanderMeer (San Francisco: Worzilla, 2010), 11.
4. For further discussion of steampunk's complicated use of the corset, see Julie Ann Taddeo, "Corsets of Steel: Steampunk's Reimagining of Victorian Femininity," in *Steaming into a Victorian Future*, ed. Julie Ann Taddeo and Cynthia J. Miller (Lanham, MD: Scarecrow Press, 2013), 43–64.
5. Lev Grossman and Gary Moskowitz, "A Handmade World," *Time* 174 (2009): 82–84.
6. Donna J. Haraway, *Simians, Cyborgs, and Women: The Reinvention of Nature* (New York: Routledge, 1991), 1.
7. Professor Calamity, "My Machine, My Comrade," *Steampunk Magazine* 3, no. 24: 25.
8. Scott Westerfeld, *Leviathan*, *Leviathan* Series (New York: Simon Pulse, 2009), 24.
9. Throughout this article, I follow the novel in using *she*, *her*, *hers* in reference to Deryn. However, it is important to note that nonbinary-identified people use a variety of pronouns.
10. Westerfeld, *Leviathan* 103.
11. Gender dysphoria is the mental distress that that transgender and gender nonconforming people experience due to lack of alignment between the gender people assign to them and their actual gender. "Gender Dysphoria." DSMV-5. American Psychiatric Association. 203. Web. 27 March 2015. http://www.dsm5.org/documents/gender%20dysphoria%20fact%20sheet.pdf
12. J. Jack Halberstam, *Female Masculinity* (Durham, NC: Duke UP, 1998), 1.
13. Westerfeld, *Leviathan* 327.

14. For further discussion of how steampunk technology pushes at the boundaries of the living and nonliving, see Lisa Hager, "Aiming to Misbehave at the Boundary between the Human and the Machine: The Queer Steampunk Ecology of Joss Whedon's *Firefly* and *Serenity*," in *The Philosophy of Joss Whedon*, ed. Dean Kowalski and Evan Kreider (Lexington, KY: UP of Kentucky, 2011), 182–93.
15. Westerfeld, *Leviathan* 149.
16. Scott Westerfeld, *Goliath*, *Leviathan* Series (New York: Simon Pulse, 2011), 379.
17. Ibid.
18. Scott Westerfeld, *Behemoth*, *Leviathan* Series (New York: Simon Pulse, 2010), 230.
19. Westerfeld, *Goliath* 329.
20. Westerfeld, *Leviathan* 383.
21. Westerfeld, *Behemoth* 244.
22. Westerfeld, *Goliath* 465.
23. Westerfeld, *Behemoth* 383.
24. Westerfeld, *Goliath* 140.
25. Ibid., 141.
26. Julia Serano, *Whipping Girl: A Transsexual Woman on Sexism and the Scapegoating of Femininity* (Emeryville, CA: Seal Press, 2007), 14–15.
27. Westerfeld, *Goliath* 145.
28. Ibid., 382.
29. Lesbian, gay, bisexual, transgender, queer/questioning, intersex, asexual.
30. Westerfeld, *Goliath* 533.
31. Scott Westerfeld, "Bonus Goliath Chapter and Art!" Scott Westerfeld, December 16, 2011, accessed August 20, 2014, http://scottwesterfeld.com/blog/2011/12/bonus-goliath-chapter-and-art/.
32. Ibid.
33. Ibid.
34. Ibid.
35. Ibid.
36. Ibid.
37. Ibid.
38. Ibid.
39. Ibid.

CHAPTER 11

STRANGE FASCINATION: KIPLING, BENJAMIN, AND EARLY CINEMA

Christopher Keep

The things I look at see me just as much as I see them.

—Valéry

N.B. The images associated with this chapter are housed in the digital annex at www.virtualvictorians.org.

Published less than a decade after the Lumière brothers mounted the first public exhibition of the cinematograph, Rudyard Kipling's short story "Mrs. Bathurst" (1904) offers one of the earliest and most troubling accounts of the seductive allure of the virtual—the immaterial realm disclosed by projecting an image onto a reflective surface, such as a screen in the case of the cinematograph, or by illuminating the pixels of a liquid crystal display in a mobile device or computer monitor. As Slavoj Žižek writes, the "virtual," in this sense, is not simply a product of our increasingly sophisticated means for projecting images; it is also and perhaps more importantly a distinctive fantasy, one that has a history stretching from "prehistoric Lascaux paintings to computer-generated Virtual Reality."[1] What defines this fantasy is the presence of a frame that offers, in its most elementary form, a way of distinguishing between self and other, between where one is and where one is not (and cannot be). "Is not the *interface* of the computer," Žižek writes, "the last materialization of this frame? What defines the properly 'human dimension' is the presence of a screen, a frame through which we communicate with the 'suprasensible' virtual universe to be found nowhere in reality."[2] Where many theorists—from psychoanalytic film scholars of the "suture school" to contemporary theorists of cyberculture—have tended to understand the fascination of the screen in terms of a narcissistic identification with the image it makes present, and in terms of the libidinal pleasures of mastery that such an identification affords, Kipling's story suggests something altogether more radical.[3] Writing at a time when cinema was not yet dominated by the machinery of realist narrative, when it still retained its earlier connection with the cinema of attractions and the scientific study of the body in motion, "Mrs. Bathurst" asks us to consider the possibility that what draws us to the virtual is not the *viewer's* narcissism, not the subject's desire to experience the ontological fullness that it otherwise lacks by way of its identification with the image on the screen. It is, rather, the narcissism of *the image itself.* That is to say, if the virtual exercises a kind of enchantment, if it arrests the viewer's gaze such that he or she seems unable to look *elsewhere*, it does so precisely by refusing to *look back*.

The fascination of the image projected on the screen might, then, be considered analogous to that which Freud ascribes to those who have come to take themselves as the object of desire and thereby do not seek love so much as they seek to be loved, to be that which the subject beholds but never possesses. Like certain types of women, but also children, animals, and some "master" criminals, the virtual holds its viewer in its thrall precisely in its "self-sufficiency and inaccessibility." "It is as if," Freud postulates, "we envied them their power of retaining a blissful state of mind—an unassailable libido-position which we ourselves have since abandoned."[4] In what follows, then, I want to read the story of one man's fascination with an image on the screen (that of Kipling's doomed warrant officer, Mr. Vickery), in the context of cinema's genealogical connections with the nonnarrative forms that dominated the period 1895–1905. In so doing, I hope to offer a theory of the virtual that moves beyond the psychodynamics of identification, the long-standing critical interest in the role of images in the constitution of the individual's ego. Instead of understanding the virtual as a mirror—that is to say, as a means of generously giving back to the subject that which it otherwise lacks—I will describe the virtual in terms of its absolute refusal to accommodate our desires even as it articulates them for us. Drawing on Walter Benjamin's writings on the "optical unconscious," I will suggest that the screen, especially as it functions in the context of early cinema, serves to frame a virtual realm from which the viewer is both perceptually and psychologically barred—and in being barred, comes all the more to desire.

THE ARRIVAL OF THE MAIL TRAIN

Kipling's "Mrs. Bathurst" is, first and foremost, a frame narrative. Like the tales of Scheherazade collected in *One Thousand and One Nights* (first translated into English in 1706) or Conrad's *Heart of Darkness* (1899), it is not only constructed as a series of stories within stories, but it also makes this arrangement integral to the narrative itself. The story concerns four men taking refuge from the heat in a railway car sitting on a siding near Cape Town, South Africa. Our narrator and his friend, a railway engineer named Hooper who has made the car available for a jaunt down the coast, are joined midway through their lunch by a naval petty officer named Pyecroft and his friend, Pritchard. The conversation winds this way and that until it comes to alight on the question of desertion, or more precisely on what a psychoanalyst would call the "object cause" of desertion, the image or ideal that would make a man risk court-martial and even execution in order to pursue some quixotic destiny beyond military service. All the men present have known someone, at some time, who has been struck by this powerful but self-destructive compulsion or urge. "It takes 'em at all ages," notes Pyecroft, who recollects a sailor named Vickery, a "superior" man who was only 18 months short of his pension when he simply disappeared following a routine expedition to secure some ammunition supplies in Bloemfontaine.[5] What, the men wonder, could compel such a person to bring dishonor to his family and throw away a secure financial future, let alone abandon his duty to Queen and country? Put another way, Kipling asks us to consider not the subject's constitution within the symbolic order, the process by which we come to assume the role of self-knowing agents within the assumed norms of the social world, so much as the dissolution of this process. What destroys rather than makes one's bond to the social?

The answer is not, as one might expect from such tales of imperial adventure, a woman, but rather the *avatar* of a woman, the virtual image that resides only in the realm of perception that is constituted by the frame of a screen, on the other side of the line that separates the human from the inhuman. On a shore leave in Cape Town some months prior to the conversation the four men are having on the railway siding, Vickery had gone to see Phyllis's Circus, a traveling show of games, exhibitions, and the latest amusements, which included the Lumières' cinematograph. As Jonathan Crary, Anne Friedberg, Lynda Nead, and others have documented,[6] cinema initially appears not as a distinctive art form in its own right, with its own production studios, distribution systems, and places of exhibition, but simply as another in a long line of visual technologies that were featured in country fairs, music halls, and other popular exhibition spaces in the nineteenth century. Panoramas, dioramas, kaleidoscopes, stereoscopes, and a host of other protocinematic devices had long sought to exploit the public's seemingly insatiable appetite for new forms of perception and embodied experience. Indeed, among the well-traveled group assembled in the train car in South Africa, the cinematograph had already run its course as a novelty by the time Vickery wandered into Phyllis's Circus. Kipling writes:

> "Oh, you mean the cinematograph—the pictures of prize fights and steamers. I've seen 'em up country."
>
> "Biograph or cinematograph was what I was alludin' to. London Bridge with the omnibuses—a troopship goin' to war—marines on parade at Portsmouth an' the Plymouth Express arrivin' at Paddington."
>
> "Seen 'em all. Seen 'em all," said Hooper impatiently. (395)

Exactly what Hooper has seen, however, is worth noting. As Tom Gunning has shown, early cinema, or what he calls the "cinema of attractions," tended to feature single-camera, fixed-point-of-view studies of exotic locales and scenes of everyday life, especially those of the modern metropolis: a group of people leaving a factory, construction workers knocking down a wall, or a parade of nannies pushing baby carriages through a park. Such familiar sights were all featured in early films by the Lumière brothers, alongside filmic postcards from India, China, and South America. These pocket documentaries, or "actualities" (each typically no more than 60 seconds long owing to the amount of film that the camera/projector could accommodate), largely eschewed the conventions of realist narrative, with its reliance on the now-familiar machinery of setting, character, and plot. Narrative cinema, especially of the kind that we now associate with the experience of "going to the movies," strives to create the fiction of a free-standing and ordered fictional universe—a world that appears to exist in and of itself, and not as the effect of specific filmic techniques and technologies. The viewer is invited to enter this world through his or her identification with the agency of the characters that drive its plot forward; their apparent ontological fullness and ability to master the diegetic space serves to fill in or paste over the gap that is constitutive of identity, and thereby affords a model for the pleasures of the symbolic order. As Laura Mulvey argues, in reference to Lacan's theory of the mirror stage,

> A male movie star's glamorous characteristics are . . . not those of the erotic object of the gaze, but those of a more perfect, more complete, more powerful ideal ego conceived at the original moment of recognition in front of the mirror. The character in the story can make things happen and control events better than the subject/spectator, just as the image in the mirror was more in control of its motor co-ordination.

The result of this identification with the point of view of the movie characters, Mulvey suggests, is a "satisfying sense of omnipotence," a surplus of pleasure that secures the viewer's place *within* the diegetic world on the screen, which now stands in for and complements the subject's sense of the real.[7]

In stark contrast to the pleasures of identification, the possibility of becoming one with the figures depicted on the screen, early cinema consistently emphasizes the gap between viewer and image. Where narrative cinema rarely directly addresses the spectator, maintaining the illusion of a fourth wall that allows for the voyeuristic pleasure of looking without being seen to look, actualities tend to act like carnival barkers calling out to the crowd, demanding that the viewer pay attention to the wonders that have been captured on film for his or her viewing pleasure. As Gunning writes, "The viewer of attractions is positioned less as a spectator-in-the-text, absorbed into a fictional world, than a gawker who stands alongside, held for a moment by curiosity and amazement." The result, then, is a distinctive form of visual display, one that "appears, attracts attention, and then disappears without either developing a narrative trajectory or a coherent diegetic world."[8] Actualities, in short, offer a world that remains stubbornly and unapologetically *other*, a spectacle to be "gawked" at, wondered over, delighted in, or even rejected as the viewer measures it against his or her own lived sense of the real.

Early cinema thus interpellates the viewer by appealing not only to his or her interest in the curious alterity of the world it discloses, but also to his or her fascination with the frame in which this world is made both present and absent at the same moment. The interface of cinema is as much the focus of the viewers' attention as the image that it encompasses and defines. Viewing a series of actualities filmed in a variety of locations around the world, gathered together in a program entitled "Home and Abroad," Vickery and Pyecroft are particularly taken by one that shows passengers disembarking from a train in London:

> "Then the Western Mail came in to Paddin'ton on the big magic lantern sheet. First we saw the platform empty an' the porters standin' by. Then the engine come in, head on, an' the women in the front row jumped: she headed so straight. Then the doors opened and the passengers came out and the porters got the luggage—just like life. Only—only when any one came down too far towards us that was watchin', they walked right out o' the picture, so to speak. I was 'ighly interested, I can tell you. So were all of us. I watched an old man with a rug 'oo'd dropped a book an' was tryin' to pick it up, when quite slowly, from be'ind two porters—carryin' a little reticule an' lookin' from side to side—comes out Mrs. Bathurst. There was no mistakin' the walk in a hundred thousand. She come forward—right forward—she looked out straight at us with that blindish look which Pritch alluded to. She walked on and on till she melted out of the picture—like—like a shadow jumpin' over a candle, an' as she went I 'eard Dawson in the ticky seats be'ind sing out: 'Christ! There's Mrs. B.!'" (398)

The film that consumes Vickery's attention inevitably recalls the fabled response to the early screenings of the Lumière brothers' actuality, *L'arrivée d'un train en gare de La Ciotat*, or, as it was known in Britain and its colonies, *The Arrival of the Mail Train* (1897). In the course of its 50-second running time, a train approaches La Ciotat Station, crossing the frame diagonally from right to left, its engine and foremost passenger cars passing out of view as it seems to simultaneously speed up and slow to stop. A group of people waiting on the platform to the right of the tracks begins to move rapidly toward the cars to secure seats, while others begin to disembark and the screen goes dark. (See annex.)

While the film may appear to be little more than a study of modern life, for many of the people who crowded into cafés, salons, and circus tents around the globe to see this latest of technological marvels, the scene was not without its more disturbing aspects. In his 1896 review of the first cinematograph exhibition in Russia, Maxim Gorky writes:

> Suddenly something clicks, everything vanishes and a train appears on the screen. It speeds straight at you—watch out! It seems as though it will plunge into the darkness in which you sit, turning you into a ripped sack full of lacerated flesh and splintered bones, and crushing into dust

> and into broken fragments this hall and this building, so full of wine, music and vice. (5)

While contemporary film scholars have cast doubt on the oft-cited stories of panic-stricken audience members shrieking at the oncoming approach of the train or attempting to flee the makeshift theaters in which the Lumières' films were first shown, Gorky's account suggests that there was, nonetheless, a degree of terror that accompanied the cinematic image.[9] Adopting the second-person address ("It speeds straight at you—watch out!"), he strives to imitate the trajectory of the locomotive as it appears to pass out of the screen and into the space of the theater, as if the frame is unable to fully contain the forces of modernity it unleashes. Kipling's restaging of this scene follows along similar lines but shifts the reader's focus from the trajectory of the train to that of one of its passengers. "Mrs. B," or, to give her her full name, Mrs. Bathurst, is a woman who runs a hotel and bar in Auckland, New Zealand, and has a particular hold on the imagination of the men who have made her acquaintance. Vickery is so struck by his momentary glimpse of a woman with whom he may or may not have had a past romantic relationship (the exact nature of their connection remains unclear) that he returns to the circus each night for the next five nights, sits patiently through the first few films on the program, stares intently as the Western Mail comes into Paddington Station, and then abruptly leaves. "The evolution never varied," Pyecroft recalls. "Two shilling seats for us two; five minutes of pictures, an' perhaps forty-five seconds o' Mrs. B walking towards us with that blindish look in her eyes an' the reticule in her hand."[10] After stepping down from the carriage and searching in vain for someone or something on the platform, Mrs. Bathurst passes not out of the screen and into the theater, as Gorky fantasizes about the train, but into some space that is located neither in the theater nor on the screen. And it is this movement—that of a single woman passing out of the frame—that seems to have the strongest hold on Vickery's attention. Mrs. Bathurst "melts" beyond the warrant officer's field of view into a realm to which he has no access, calling his attention to the vanishing point around which the image is organized but that can never be directly represented on the screen, save as a gap or void. For Kipling, however, the enigma of what the frame encloses and what it fails to enclose leads not to panic, but to *fascination*. It provokes not a desire to look away, but the inability to look *elsewhere*. Vickery returns each night to watch the same image repeat exactly the same actions, again and again.

A Tiny Spark of Contingency

Kipling's account of the affective force of early cinema, its ability to hold the viewer's attention more powerfully than other technological means of representation, emphasizes the way in which such chance encounters yield, as their unexpected surplus, a moment of insight into the nature of the virtual. Like Mrs. Bathurst with her "blindish look," the Lumières' camera, set up in a fixed location and recording a single scene for the duration of 50 seconds, looks but fails to see. Its lens takes in everything within its depth of field: from the curlicues of smoke belching from the locomotive's stack to the porters assembling on the platform in anticipation of the train's arrival, no one thing is singled out and no single detail carries a greater epistemic charge than any other in the crowded fullness of the frame. And it is precisely the unmoving, unfeeling, purely mechanical nature of this gaze that allows the motion picture camera to capture something that the human eye cannot register. Siegfried Kracauer writes of the Lumières' actualities:

> Their themes were public places, with throngs of people moving in diverse directions . . . It was life at its least controllable and most unconscious moments, a jumble of transient, forever dissolving patterns accessible only to the camera. The much imitated-shot of the railway station, with its emphasis on the confusion of arrival and departure, effectively illustrated the fortuity of these patterns. (31)

If the nonnarrative actuality film differs from other forms of the virtual, from the camera obscura to the mobile computing device, it is in the way that it brings into focus the "jumble of transient, forever dissolving patterns" that make up the modern world but that would otherwise pass below the level of conscious perception. *Baby's Tea-Time* (1895), to take as our example another of the films featured in the Lumières' early program, seems simple enough: a man and a woman sit at a table situated in the family's outdoor garden, with their baby sitting in a high chair between them. The father spoons baby food from a saucer into the child's mouth; a portion drips down the child's face while the woman sips at her tea. (See annex.)

The film appears to capture a charming scene from domestic life, but it was not, as one might expect, the baby's chubby cheeks that had audiences calling for the film to be shown again and again, but something the filmmakers had not so much as noticed at the time of its production. As Bertrand Tavernier says in the audio commentary to a DVD compilation marking the hundredth anniversary of the cinematograph, "What impressed most [sic] the audience at the time was the leaves moving, the wind in the trees." Projected onto the magic lantern screen by the incandescent light of the Lumières' projector, such mundane details acquired a new kind of interest. It was not simply the indexicality of such images, the way in which they seemed to register life as it was, in and of itself, but rather their absolute alterity, the way in which they seemed to belong to a clandestine realm of fluxes and fleeting intensities of energy that existed both within and yet somehow beyond the perceived world.

Walter Benjamin describes the realm disclosed by the blindness of the film camera as "the optical unconscious." Freud's *The Psychopathology of Everyday Life* (1901), Benjamin argues, alerted readers to a whole range of human experience, from momentary acts of forgetfulness to humorous slips of the tongue, that contain clues to our psychic lives far in excess of their apparent importance (or lack thereof). Just as Freud's book "isolated and made analyzable things which had previously floated unnoticed on the broad stream of perception," so, too, Benjamin suggests, cinema has afforded "a similar deepening of apperception throughout the entire spectrum of optical—and now also auditory—impressions."[11] The analogy here between the psychical and the optical unconscious is at best imperfect. The unconscious, as Freud posits it, belongs to and within the subject; it is that interior realm that is opened up by the secondary processes as they seek to repress the desires and needs that are incompatible with social norms and are therefore a threat to the subject's ability to function within those norms. Hence, the signs it produces are those that come into view only insofar as they are translated into some more acceptable form (most typically through the action of condensation or displacement) and thereby escape the regulatory apparatus of the superego. By contrast, Benjamin's optical unconscious is a realm that belongs neither to the subject nor to the object world, but to both at once. "Clearly, it is another nature," Benjamin writes, "which speaks to the camera as compared to the eye. 'Other' above all in the sense that a space informed by human consciousness gives way to a space informed by the unconscious."[12] This "other nature," opened up by the camera's eye, is the same as that which we encounter every day but have never "seen." As Benjamin puts it in a well-known passage, "We are familiar with the movement of picking up a cigarette lighter or a spoon, but know almost nothing of what really goes on between hand and metal, and still less how this varies with different moods."[13] The cinema, in this sense, must be understood not simply as a technology, a means of recording the actions of bodies in real time, but also as a form of epistemology, a means of understanding the dialectical relation between interior and exterior, or, as Benjamin would have it, the mutually constitutive nature of "mood" and "motion."

Benjamin's sense of the virtual as a kind of "other nature," and the camera as a means of discerning its occulted structures of meaning, has deep roots in the Victorian period and its scientific studies of the body in motion. Beginning with a striking series of experiments in 1860, Etienne-Jules Marey argued that the body, whether human or animal, was but an animate machine, the actions of which were governed by the same forces that operated in the inanimate realm. There was, in his view, no special vital force immanent in organic tissue that gave it life, no "one intellectual spirit," as Coleridge put it, "plastic and vast" that "sweeps through all things." There was only the mechanical means of converting matter into energy that one saw at work in steam boilers, the terms for which had been recently set out in the first and second laws of thermodynamics. The result, as Marta Braun writes, was a "new perception of the human body":

> Where it once was imaginable as a static machine or as the locus of a metaphysical vital force, the body was now perceived as a field of energies whose forms—yet to be discovered—were conceivable in terms of the laws governing light, heat, magnetism, and electricity. (*Picturing Time* 14)

The primary obstacle that Marey faced in his efforts to discover the laws of "animal motion" was procuring precise measurements of the body's movements, most of which, he found, were effectively invisible due to their swift and fluid transitions from state to state. Unaided, the senses are "baffled alike by objects too small or too large, too near or too remote, as well as by movements too slow or too rapid."[14] In response to this problem, Marey devised what he called "chronography," a notational system whereby the body itself would produce the legible signs of its movement through time, thus leaving a permanent record that might be studied and interpreted where the body itself could not. Marey's first effort was a device for measuring the action of the heart as it contracted and expanded. The sphygmograph adapted earlier instruments for regulating steam pressure in a cylinder, substituting the motion created by the displacement of the arterial wall for that of the piston in a steam engine. A sensitive needle placed just above the pulse point in the wrist communicated the contractions and expansions to a stylus, which, in turn, converted these movements into a fluid script on a thin sheet of smoke-blackened paper; a small clockwork mechanism moved the paper along at a uniform rate so that it was possible to correlate the script to the passage of time.

The sphygmograph was an important breakthrough in the study of physiology. It did away with the vagaries of personal observation and its uneasy passage into words, emancipating the resulting data from the flights of figurative language that marred the linguistic form of narrative. Here the body itself seemed to write in its own language, providing experimenters with an apparently objective record of bodily movement from which to derive its mathematical relations and hence its scientific character. Such experiments paved the way for a whole series of protocinematic instruments for measuring aspects of bodily motion, including a pair of shoes that allowed the wearer to record the number, length, and frequency of his or her steps, as well as variations in foot pressure, all rendered as a series of curving lines traced onto smoke-blackened paper. These instruments produced, as Mark Seltzer puts it, "a sort of automatic writing that unite[s] the body's own signs (pulse, heart rate, gait, the flapping of wings) with a language of technical representation."[15]

Marey's interest in the temporality of the body in motion, and the ways in which this study yielded a sense of a realm beyond human capacity to perceive or comprehend, was shared by Eadweard Muybridge. Hired by Leland Stanford, a former governor of California and prominent businessman, to resolve the debated issue of whether or not there is a moment in a horse's stride when all four legs are free of the ground, Muybridge devised a system for taking still photographs in rapid succession. Setting up his equipment along a specially built track in Palo Alto, Muybridge arranged 12 cameras in a line before a wooden backdrop covered with a white cloth that had been marked with dark vertical lines 21 inches apart. Each of the spaces between the lines was numbered from 1 to 20, in order to indicate the horse's progress. At the bottom of this screen was another that was measured off with a series of horizontal lines, each 4 inches apart, in order to indicate the height of the horse's legs above the ground. Electrical wires were run from each camera under the track so that they corresponded with each of the vertical lines on the rear screen. A portion of each wire was exposed at just the point where one of the wheels of the sulky would come into contact with it, effectively breaking the circuit and causing the shutter to expose the lens. Before an invitation-only crowd on June 15, 1878, Stanford's prize trotter, Abe Edgington, was sent down the track. The cameras clicked in rapid succession. The results of the photographs, which Muybridge developed while the assembled crowd waited, successfully resolved the "unsupported transit controversy"—one of the images clearly showed that, at one point in his stride, Abe Edgington was indeed airborne. (See annex.)

Learning of these experiments across the Atlantic, Marey was ecstatic. Imagining "a kind of *photographic* gun" that would allow its bearer to capture the movements of a bird's wings in full flight, he writes:

> It is clearly an easy experiment for Mr. Muybridge. And then what beautiful zoetropes he will be able to give us: in them we will see all imaginable animals in their true paces; it will animate zoology. As for artists, it is a revolution, since they will be provided with the true attitudes of movement, those positions of the body in unstable balance for which no model can *pose*. (qtd. in Braun, *Muybridge* 152)

However, Marey's enthusiasm for the way in which Muybridge's photographs allowed viewers to see, as if for the first time, the "true paces" of the animal machine was not universally shared. While many did express great excitement on seeing the flickering images of the horse, quite a few artists, those for whom Marey had thought that moving pictures would effect a kind of "revolution," were deeply disturbed by what the camera revealed. When time was divided into mere fractions of a second, the much-admired grace and fluidity of the horse's movements were shown to be made up of a series of awkward gestures that were worryingly reminiscent of the twitchings and spasms of a fit or a convulsion, as if the ugliness associated with sickness were somehow inscribed within the beauty of health. The contrast between the new virtual reality and that which had been afforded by the unaided eye was especially disturbing to artists who had made "truth to nature" the highest measure of the aesthetic worth of a work—a relatively common stance in the nineteenth century. Responding to Muybridge's images in the *Gazette des Beaux-Arts*, Georges Gueroult writes, "The attitudes are, for the most part, not only ungraceful, but have a false and impossible appearance." As such they cannot be true to nature, for nature is as we see it, and artists would be best advised to abjure depending on such sources for their own work; they should, he urges, "speak their own language."[16] Even such an eminent observer of the body as Rodin felt the need to warn his fellow artists against attending to this "other nature":

> It is the artist who is truthful and it is photography which lies, for in reality time does not stop, and if the artist succeeds in producing the

> impression of a movement which takes several moments [to accomplish], his work is certainly much less conventional than the scientific image, where time is abruptly suspended. (91)

The realm of perception opened up by these chronophotographic experiments appeared not only to draw attention to the limits of the human eye; it also, and at the same time, had the curious effect of making nature itself seem *unnatural.* The virtual thus appears in these debates concerning animal motion not as the mirror image of the real, but as a distinctive realm in its own right; it may offer the "truth," but its truth, captured through the unfeeling eye of the camera, was somehow inhuman.

Benjamin's dialectical conception of the optical unconscious, however, encourages us to consider both the alterity of the virtual, the degree to which it refuses to give back to the viewer his or her sense of the world, and the elusive forms of knowledge afforded by this refusal. For Benjamin, cinema is a frame that mediates the experience of this hidden dimension and the unconscious desires of the observer. And the key to this experience is the element of *chance*, the possibility that the camera, with its "blindish look," will record something other than what its operator intended to capture—something, indeed, beyond the capacity of the human eye to register in ordinary circumstances. The image, recorded at some time in the past and wholly independent of the observer's circumstances, seems somehow intimately connected with his or her own storehouse of mental images. Thus it opens the self to itself, or rather to that part of the self from which it has been alienated. The image, in this way, seems so perfectly attuned to our inner lives, so deeply aware of our most intimate fears and desires, that it appears *to have been made for us.* According to Miriam Hansen, "The optical unconscious thus as much refers to the psychic projection and involuntary memory triggered in the beholder as it assumes something encrypted in the image that nobody was aware of at the time of exposure."[17] Benjamin calls such chance elements "the tiny spark of contingency," or "the mark of the here and now" that characterizes every photographic image (insofar as it bears an indexical relation to the scene it records at a specific moment in time) but that can only be "seen" at some point in the future, the moment at which it ceases to be incidental detail and suddenly becomes meaningful to the viewer.[18]

The aleatory "spark" that prompts the encounter of the self with itself is key to understanding the allure of cinema. In Kipling's "Mrs. Bathurst," to return to our tale of cinematic fixation, what brings

Vickery back to Phyllis's Circus every night is not simply the opportunity to see a train pull into Paddington Station, nor even the strangeness of witnessing a familiar face magically transported to the screen. It is, rather, the way in which this chance encounter appears to recall and remake (or perhaps unmake) something in himself. One night, Pyecroft, who has taken to accompanying Vickery on his nightly sojourns to the circus in the hopes of cadging a drink from him afterward, says to the troubled man, "I wonder what she's doin' in England? . . . Don't it seem to you she's lookin' for somebody?" Kipling describes the rest of the scene with care:

> That was in the Gardens again, with the South-Easter blowin' as we were makin' our desperate round. "She's lookin' for me," he says, stoppin' dead under a lamp an' clickin' 'is four false teeth like a Marconi ticker. "Yes! lookin' for me," he said, an' he went on very softly an' as you might say affectionately. "*But,*" he went on, "in future, Mr. Pyecroft, I should take it kindly of you if you'd confine your remarks to the drinks set before you. Otherwise," he says, "with the best will in the world towards you, I may find myself guilty of murder!" (401)

The chance sighting of Mrs. Bathurst on the platform at Paddington Station is the "tiny spark of contingency" around which the text is organized. That is to say, something that is only incidental to the film's ostensible subject matter (the arrival of the train) provokes Vickery to contemplate something within himself of which he appears to have been previously unaware, but which is serious enough to prompt him to threaten the life of his fellow serviceman. What exactly this private revelation is, however, remains unclear. Indeed, as Nicholas Daly writes in his overview of the story's critical reception, literary scholars who have addressed the text have often found it inscrutable.[19] Kingsley Amis described it as "unintelligible," and Angus Wilson, going a step further, dismissed it as "pretentious." "The 'difficulty' of 'Mrs. Bathurst,'" he writes, "is of little interest, for in the last resort, the story is empty."[20] But this "emptiness" is, in fact, the point. Kipling's refusal to fill in many of the narrative details that would allow us to better understand Vickery's relation to Mrs. Bathurst brings us to the heart of the virtual as a form of knowledge. It is the very fact that Vickery cannot truly know whom or what Mrs. Bathurst is looking for as she steps down from the train that allows him to assume she has come to London in search of him. What we might call the missing addressee of her look, the fact that it lacks any identifiable object, serves to make "visible" that which cannot be directly represented—that is, the vanishing point around which the image within the frame is organized.

This evocation of a gap or lack in the image—that which exceeds the viewer's capacity to know—seems, in turn, to bring Vickery into some proximity with his own lack. He confronts the psychical emptiness that not only characterizes his life as a warrant officer in the service of Queen and country (although he seems to have taken great pride in this role) but that is also constitutive of identity itself. Shortly after Phyllis's Circus leaves town and he is no longer able to return to the site of his traumatic encounter with the lack that defines the real, he volunteers to lead an exhibition inland to retrieve a supply of ammunition left behind following the cessation of hostilities (the story appears to take place shortly after the Boer War). He then . . . disappears. "Went out—deserted, if you care to put it so," says Pyecroft, "within eighteen months of his pension."[21] Just as Mrs. Bathurst "melts" from the frame of the cinematograph, so, too, Vickery slips beyond the field of vision of Kipling's elaborately constructed frame narrative.

THE PETRIFIED FOREST

For Benjamin, the spark of contingency that resides at the heart of the cinematic image provides grounds for hope. The momentary shock produced by such chance encounters of the self with itself might simply reproduce the anaesthetizing effect of modern life, the way in which its discontinuities and brutal assaults on the senses serve to accustom the subject to its alienation from the means of production. But it might also serve as a means to break the subject out of its torpor and to create the space in which it will be possible to imagine new ways of organizing the socius. As Benjamin writes in his defense of Sergei Eisenstein's *Battleship Potemkin* (1925):

> To put it in a nutshell, film is the only prism in which the immediate environment—the spaces in which people live, pursue their avocations, and enjoy their leisure—are laid open before their eyes in a comprehensible, meaningful, and passionate way. In themselves these offices, furnished rooms, bars, big-city streets, stations, and factories are ugly, incomprehensible and hopelessly sad. Or rather they were and seemed to be, until the advent of film. The cinema then exploded this prison-world with the dynamite of its fractions of a second, so that now we can set off calmly on journeys of adventure among its scattered ruins. ("Reply to Oscar A. H. Schmitz" 218)

Precisely because it is made up of a sequence of still photographs, imposing itself on the eye in such rapid succession that the subject can never come to terms with any one before another replaces it, cinema offers a space in which the subject might learn to manage the otherwise destructive rhythms and modalities of modern life. As Hansen helpfully glosses this passage, "By refracting the modern *physis*, film simultaneously transforms it: 'With the dynamite of the split second,' it denaturalizes the entire 'prison-world,' undoes its semblance of immutability, and makes its scattered ruins available for mimetic transformation and reconfiguration."[22] Cinema, in this instance, is both the fullest manifestation of and the surest way to overcome the enchantments of what Mark Fisher calls "capitalist realism."

Kipling was less optimistic. Where many film theorists, as I have demonstrated, have tended to see cinema as answerable to the pleasure principle, and hence as a model for the psychical process whereby the subject takes up its place within the social world, "Mrs. Bathurst" aligns the medium more strongly with the death drive. Just as Freud's studies of a child who seems to compulsively rehearse the disappearance of its mother led him to posit the existence of a fundamentally conservative force, one that seeks "to restore an earlier state of things which the living entity has been obliged to abandon under the pressure of external disturbing forces,"[23] so, too, Vickery's repeated viewings of Mrs. Bathurst as she passes into and out of the frame lead him to reject the symbolic order and the forms of deferred gratification it affords. As Pyecroft's narration of his time with Vickery draws to a close, Hooper, the engineer who has made this sidecar available in the first place, returns to a story he had begun to tell earlier. It concerns his own trip into the interior to look for a pair of tramps who had disappeared into a peculiarly dense section of teak forest, "a sort of mahogany really," that runs some 70 miles without a bend beyond Buluwayo.[24] Another account of railway travel and the forms of perception it affords follows, but in this case, the point of view is not that of the person on the platform watching as the train pulls into the station, but that of the train itself as it pushes into the darkness:

> "We get heaps of tramps up there since the war. The inspector told me I'd find 'em at M'Bindwe siding waiting to go North. . . . I went up on a construction train. I looked out for 'em. I saw them miles ahead along the straight, waiting in the teak. One of 'em was standin' up by the dead-end of the siding an' the other was squattin' down lookin' up at 'im, you see."
>
> "What did you do for 'em?" said Pritchard.
>
> "There wasn't much I could do, except bury 'em. There'd been a bit of a thunderstorm in the teak, you see, and they were both stone dead and as black as charcoal. That's what they really were, you see—charcoal. They fell to bits when we tried to shift 'em." (407)

The straight run of track through the forest effectively simulates the seamlessly shifting perspective provided by the motion picture camera as it dollies into a scene.[25] And, perhaps not surprisingly, it too affords a moment of uncanny revelation. What appears to be two men idly waiting along the railway siding is revealed to be, as the train approaches more closely, a kind of mausoleum where the dead have been interred in the poses of the living. All that remains of Vickery, who appears to have been the standing figure in the tableau that Hooper describes, are his false teeth and the remnants of a tattoo, now rendered as a "negative" image, white on black, by the intensity of the fire.

Vickery's cinematic fixations lead, then, not to the "dynamiting" of the "prison-house" that holds the modern subject in perceptual chains, but to his transformation into mere *matter*, the state of zero excitation, the null point from which life began. As Freud puts it in *Beyond the Pleasure Principle*, "If we are to take it as a truth that knows no exception that everything living dies for *internal* reasons—becomes inorganic once again—then we shall be compelled to say that '*the aim of all life is death*' and, looking backwards, that '*inanimate things existed before living ones.*'"[26] The death drive, in this sense, is the desire to end the perpetual play of signification that drives the subject further and further into the symbolic order, ever searching for that which will seal up, once and for all, the gap at the heart of being. It is *the desire to end desire.*

Such is the final insight afforded by Kipling's study of the affective force of cinema: the fascination of the virtual realm, the way in which it compels us to return, over and over again, to the seemingly insurmountable fact of the screen, stems not from our technological mastery of the natural world, not from the narcissistic desire to behold again what humans have accomplished (and, indeed, what they might in future continue to accomplish) with their tools. It stems, instead, from a certain kind of constitutive *failure*, the inability to master that which exceeds the bounds of the symbolic order itself, the surplus that, much like Mrs. Bathurst in the actuality shown at Phyllis's Circus, perpetually melts from view. As Žižek writes, "This hole [in the interface between self and image] derails the balance of our embededdness in the natural environs and throws us into the state of 'out-of-joint': no longer 'at home,' striving for the Other Scene which, however, remains forever 'virtual,' a promise of itself, a fleeting anamorphic glimmer accessible only to a side view."[27] What holds us to the screen is not, then, our narcissism, but the narcissism of the image. As the late-Victorian study of kinetics and the optical unconscious demonstrates, the virtual is defined not by its verisimilitude but by its apparent alterity, the way it seems to refuse to conform to human expectations of what is and is not "real," what is and is not "natural." Seemingly complete and whole in a way that the observing subject is not and cannot be, the virtual perpetually withdraws from the gaze of the viewer, indifferent to his or her desires, needing only itself. Kipling's ill-fated warrant officer, with his clicking false teeth and insatiable appetite for moving pictures, may die thinking, believing, hoping that Mrs. Bathurst is traveling the world, looking for him, that *he* is the object cause of *her* desire. But this is simply a fantasy, albeit one that so many have come to share in this age of compulsive connectivity; it is the necessary illusion, made possible by the vanishing point around which the image is organized, that the other turns its gaze to me and to me alone. Just who or what the avatar truly desires—if it desires at all—necessarily remains on the other side of the screen.

NOTES

1. Žižek 98.
2. Ibid.
3. On the "suture school" of psychoanalytic film scholarship and the role of narcissism in the experience of cinematic spectatorship, see Christian Metz, "The Imaginary Signifier," *Screen* 16, no. 2 (1975): 14–76; Jacques Alain Miller, "Suture (Elements of the Logic of the Signifier)," *Screen* 18, no. 4 (1977–78): 21–25; Laura Mulvey, "Visual Pleasure and Narrative Cinema," *Screen* 16, no. 3 (1975): 6–18; and Kaja Silverman, *The Subject of Semiotics* (Oxford: Oxford UP, 1983). On the function of narcissism in computer-mediated communications and virtual culture, see Andre Nusselder, *Interface Fantasy: A Lacanian Cyborg Ontology* (Cambridge, MA: MIT Press, 2009); Claudia Springer, *Electronic Eros: Bodies and Desire in the Postindustrial Age* (Austin: University of Texas Press, 1999); and Sherry Turkle, *Life on the Screen: Identity in the Age of the Internet* (New York: Simon and Schuster, 1995).
4. "On Narcissism" 46.
5. Kipling 386, 388.
6. On the connections between cinema and the exhibitionary culture of the nineteenth century, see Jonathan Crary, *Techniques of the Observer: On Vision and Modernity in the Nineteenth Century* (Cambridge: Cambridge UP, 1990); Miriam Bratu Hansen, *Babel and Babylon: Spectatorship in American Silent Film* (Cambridge, MA: Harvard UP, 1991); Anne Friedberg, *Window Shopping: Cinema and the Postmodern* (Berkeley: University of California Press, 1993); and Lynda Nead, *The Haunted Gallery: Painting, Photography, Film, c. 1900* (New Haven, CT: Yale UP, 2007).
7. Mulvey 12.

8. "The Whole Town" 190, 193.
9. While there are no firsthand reports of spectators openly panicking at the sight of the oncoming train when the Lumières' films were first exhibited, the fleeting fantasy that it might somehow pass into the theater itself is a commonplace of many contemporary reviews, including (in addition to Gorky's) those of Felix Renault (in France) and Ottomar Volkmer (in Austria). See Martin Loiperdinger, "Lumière's 'Arrival of the Train': Cinema's Founding Myth," *The Moving Image* 4, no. 1 (2004): 89–118.
10. Kipling 400.
11. "Work of Art" 265.
12. Ibid., 266.
13. Ibid.
14. Marey 286.
15. Seltzer 178.
16. Qtd. in Forster Hahn 105.
17. Hansen 156.
18. "Little History of Photography" 510.
19. Daly 68.
20. Wilson 223.
21. Kipling 405.
22. Hansen 159.
23. *Beyond the Pleasure Principle* 30.
24. Kipling 406.
25. According to Lynne Kirby, the train might be thought of as "cinema's mirror image in the sequential unfolding of a chain of essentially still images and the rapid shifts of point of view that the train and cinema experiences entail." See her *Parallel Tracks: The Railroad and Silent Cinema* (Durham, NC: Duke UP, 1997), 2.
26. *Beyond the Pleasure Principle* 32.
27. Žižek 99.

SELECT BIBLIOGRAPHY

#TransformDH. Last modified May 5, 2014. Accessed May 5, 2014. http://www.transformDH.org.

Aarseth, Espen J. *Cybertext: Perspectives on Ergodic Literature*. Baltimore: John Hopkins UP, 1997.

The Aberdeen Journal, June 30, 1830.

Anon. *A Picturesque Guide to the Regent's Park: With . . . Descriptions of the Colosseum, the Diorama, and the Zoological Gardens*. London: John Limbird, 1829.

Anon. "Autobiography of Leigh Hunt," *Tait's Edinburgh Magazine*, 17 (September 1850): 563–72.

Anon. "Leigh Hunt and His Contemporaries," *Literary Gazette and Journal of Belles Lettres*, 1745 (June 29, 1850): 436–37.

Anon. "Leigh Hunt," *Dublin University Magazine*, 36 (September 1850): 268–86.

Anon. "Leigh Hunt's *Autobiography*," *Spectator* 23 (June 22, 1850): 593–94.

Anon. "Review of *Autobiography*," *Palladium*, 1 (August 1850) 130–39.

Anon. "Review of Leigh Hunt's *Autobiography*," *Chamber's Edinburgh Magazine*, 14 (July 13, 1850): 19–23.

Anon. "Review of *The Autobiography of Leigh Hunt*," *Literary World*, 7 (September 14, 1850): 209–10.

Anon. "Some Account of the Colosseum, in the Regent's Park." *The Mirror of Literature, Amusement, and Instruction* 13, no. 352, January 17, 1829, 34–37.

Anon. "*The Autobiography of Leigh Hunt*," *Harper's New Monthly Magazine* 14 (November 1850): 143–68.

Anon. "*The Autobiography of Leigh Hunt*," *North British Review* 14 (November 1850): 143–68.

Anon. "The Autobiography of Leigh Hunt," *The Times* 20585 (September 4, 1850): 7.

Anon. "The Colosseum." *London Magazine*, February, 1829: 104–8.

Archer, Mildred, and Ronald Lightbown. *India Observed: India as Viewed by British Artists 1760–1860*. London: Victoria and Albert Museum, 1982.

Armstrong, Nancy. "Realism before and after Photography: The Fantastical Form of a Relation among Things." In *A Concise Companion to Realism*, edited by Matthew Beaumont, 102–20. Oxford: Wiley-Blackwell, 2010.

Armstrong, Nancy, and Leonard Tennenhouse. "A Mind for Passion: Locke and Hutcheson on Desire." In *Politics and the Passions, 1500–1850*, edited by Victoria Kahn, Neil Saccamano, and Daniela Coli, 131–50. Princeton, NJ: Princeton UP, 2006.

Arnott, Neil. *Elements of Physics, or Natural Philosophy*. 2 vols. London: Longman, Rees, Orme, Brown, and Green, 1829.

Arthur, Paul. *Virtual Voyages: Travel Writing and the Antipodes 1605–1837*. London: Anthem Press, 2010.

Atkinson, Juliette. "Fin-de-Siècle Female Biographers and the Reconsideration of Popular Women Writers." In *Writing Women of the Fin de Siècle: Authors of Change*, edited by Adrienne E. Gavin, Carolyn W. de la L. Oulton, and Linda H. Peterson, 111–23. New York: Palgrave Macmillan, 2012.

Bacon, Francis. *De Augmentis Scientiarum*. In *The Philosophical Works of Francis Bacon*, edited by John M. Robertson. 1905; rpt. Abingdon, UK: Routledge 2011.

Baker, Paul, and Amanda Potts. "'Why Do White People Have Thin Lips?' Google and the Perpetuation of Stereotypes via Auto-complete Search Forms." *Critical Discourse Studies* 10, no. 2 (2013): 187–204.

Barnett, Harry V. "The Special Artist." *The Magazine of Art*. London: Cassell and Company, 1883.

Bastian, Mathieu, Sebastien Heymann, and Mathieu Jacomy. "Gephi: An Open Source Software for Exploring and Manipulating Networks." *ICWSM* 8 (2009): 361–62.

Baudelaire, Charles. *The Painter of Modern Life and Other Essays*. Translated and edited by John Mayne. London: Phaidon Press, 1964.

Beetham, Margaret. *A Magazine of Her Own? Domesticity and Desire in the Woman's Magazine, 1800–1914*. London: Routledge, 1996.

Bell, Leonard. "To See or Not to See: Conflicting Eyes in the Travel Art of Augustus Earle." In *Orientalism Transposed: The Impact of the Colonies on British Culture*. Edited by Julie F. Codell and Dianne Sachko Macloed, 117–39. Aldershot, UK: Ashgate, 1998.

Benjamin, Walter. "Little History of Photography." *Walter Benjamin: Selected Writings*. Edited by Michael W. Jennings and Howard Eiland. Cambridge, MA: Belknap, 1999–2003. 4 vols. 2:507–30.

———. "Reply to Oscar A. H. Schmitz." *Walter Benjamin: Selected Writings*. Edited by Michael W. Jennings and Howard Eiland. Cambridge, MA: Belknap, 1999–2003. 4 vols. 2:16–19.

———. "The Work of Art in the Age of Its Technological Reproducibility: Third Version." *Walter Benjamin: Selected Writings*. Edited by Michael W. Jennings and Howard Eiland. Cambridge, MA: Belknap, 1999–2003. 4 vols. 4:251–83.

Bennett, Geoffrey. *Charlie B: A biography of Admiral Lord Beresford of Metemmeh and Carraghmore*. London: Peter Dawnay, 1968.

Berners-Lee, Tim, James Hendler, and Ora Lassila. "The Semantic Web." *Scientific American*, May 17, 2001. Accessed April 28, 2014. http:// www .scientificamerican.com/article/the-semantic-web/.

Besser, Howard. "The Past, Present, and Future of Digital Libraries." In *A Companion to Digital Humanities*. Edited by Susan Schreibman, Ray Siemens, and John Unsworth. Blackwell, 2004.

Blei, David M. "Topic Modeling and Digital Humanities." *Journal of Digital Humanities* 2, no. 1 (2012). Accessed April 28, 2014. http://journalofdigitalhumanities.org/2-1/topic- modeling-and-digital-humanities-by-david-m-blei/.

Blondel, Vincent D., Jean-Loup Guillaume, Renaud Lambiotte, and Etienne Lefebvre. "Fast Unfolding of Communities in Large Networks." *Journal of Statistical Mechanics: Theory and Experiment* 2008, no. 10 (2008): P10008. doi:10.1088/1742-5468/2008/10/P10008.

Bode, Katherine. *Reading by Numbers: Recalibrating the Literary Field*. London: Anthem Press, 2012.

Boellstorff, Tom. *Coming of Age in Second Life: An Anthropologist Explores the Virtually Human*. Princeton, NJ: Princeton UP, 2008.

Booth, Alison. Collective Biographies of Women. Online bibliography. 2013. http://womensbios.lib.virginia.edu. Database. 2013. http://cbw.iath.virginia.edu/public/index.php.

———. *How to Make It as a Woman: Collective Biographical History from Victoria to the Present*. Chicago: University of Chicago Press, 2004.

———. "Prosopography." In *The Wiley-Blackwell Encyclopedia of Victorian Literature*, forthcoming.

Bourdieu, Pierre. "The Field of Cultural Production, or: The Economic World Reversed." In *The Field of Cultural Production: Essays on Art and Literature*. Edited by Randal Johnson, 29–73. New York: Columbia UP, 1993.

———. *The Rules of Art: Genesis and Structure of the Literary Field*. Translated by Susan Emanuel. Stanford, CA: Stanford UP, 1996.

Bowers, Fredson. *Principles of Bibliographic Description*. Princeton, NJ: Princeton UP, 1949.

Brake, Laurel. *Subjugated Knowledges: Journalism, Gender and Literature in the Nineteenth Century*. Houndmills, UK: Macmillan, 1994.

Braun, Marta. *Picturing Time: The Work of Étienne-Jules Marey (1830–1904)*. Chicago: University of Chicago Press, 1992.

———. *Eadweard Muybridge*. London: Reaktion, 2010.

Bristow, Joseph, ed. *The Fin-de-Siècle Poem: English Literary Culture and the 1890s* Athens: Ohio UP, 1995.

Britton, John. *A Brief Account of the Colosseum, in the Regent's Park*. London, 1829.

Britton, John, and Augustus Pugin. *Illustrations of the Public Buildings of London*. 2 vols. London: J. Taylor, 1825, 1828.

Brown, Susan, and John Simpson. "The Curious Identity of Michael Field and Its Implications for Humanities Research with the Semantic Web." Paper presented at the IEEE Conference on Big Data, Santa Clara, California, October 6–9, 2013. Accessed April 28, 2014. doi:10.1109/BigData.2013.6691674.

Brown, Susan, Patricia Clements, and Isobel Grundy. *Orlando: Women's Writing in the British Isles from the Beginnings to the Present*. Cambridge: Cambridge UP, 2006–14. Accessed April 28, 2014. http://orlando.cambridge.org.

Brown, Susan, Patricia Clements, and Isobel Grundy. "Sorting Things In: Feminist Knowledge Representation and Changing Modes of Scholarly Production." *Women's Studies International Forum* 29, no. 3 (2006): 317–325.

Bugajski, Ken A. "Editing and Noting: Vision and Revisions of Leigh Hunt's Literary Lives." *Romanticism and Victorianism on the Net* 50 (2008).

Burlingame, Michael. *Abraham Lincoln: A Life*. Baltimore: Johns Hopkins UP, 2008.

Byatt, A. S. *Possession*. London: Chatto & Windus, 1990.

Byerly, Alison. *Are We There Yet? Virtual Travel and Victorian Realism*. Ann Arbor: The University of Michigan Press, 2013.

Calamity, Professor. "My Machine, My Comrade." *Steampunk Magazine* 3, no. 24 (2007): 24–25.

Cameron, K. N. (ed.). *Shelley and His Circle, 1773–1822*, vols 1–4. Cambridge, MA: Harvard UP, 1961–70.

Carlyle, Thomas. *The Collected Letters of Thomas and Jane Welsh Carlyle, Vol. 25 – 1850*, eds. Ian Campbell, Aileen Christianson, and Hilary J. Smith. Durham and London: Duke UP, 1997.

———. "Thoughts on History." *Fraser's Magazine for Town and Country* 2, no. 10 (November 1830): 413–18.

Chapman, Alison, and Caley Ehnes, eds. *Victorian Periodical Poetry*, special issue of *Victorian Poetry* 52, no. 1 (Spring 2014).

Clayworth, Anya. "*The Woman's World*: Oscar Wilde as Editor." *Victorian Periodicals Review* 30, no. 2 (Summer 1997): 84–101.

Codell, Julie F., ed. *Imperial Co-Histories—National Identities and the British Colonial Press*. Madison, UK: Fairleigh Dickinson UP, 2003.

Cohen, Dan. "A Conversation with Data: Prospecting Victorian Words and Ideas." Dan Cohen. http://www.dancohen.org/page/2/.

Cohen, Michael. "E. C. Stedman and the Invention of Victorian Poetry." *Victorian Poetry* 43, no. 2 (Summer 2005): 165–88.

Colosseum, Regent's Park (London). *A Description of the Royal Colosseum Re-opened in 1845, Re-embellished in 1848*. Twenty-second edition. London: Printed by J. Chisman, 1848.

Comment, Bernard. *The Panorama*. London: Reaktion Books, 1999.

comScore, Inc. "comScore Releases July 2013 U.S. Search Engine Rankings." comScore, Inc. Accessed October 23, 2013. http://www.comscore.com/Insights/Press_Releases/2013/8/comScore_Releases_July_2013_U.S._Search_Engine_Rankings.

Cope, Bill, and Mary Kalantzis. "The Role of the Internet in Changing Knowledge Ecologies." *Arbor* 737 (2009): 521–30.

Cox, Jeffrey N., and Greg Kucich. "Introduction." In *The Selected Writings of Leigh Hunt*, Edited by M. Eberle-Sinatra and R. Morrison, 6 vols. London: Pickering & Chatto, 2003.

Crane, Gregory. "What Do You Do with a Million Books?" *D-Lib Magazine* 12, no. 3 (March 2006). Accessed April 28, 2014. doi:10.1045/march2006-crane.

Cronin, Richard. *Reading Victorian Poetry*. Oxford: John Wiley, 2012.

———. *Romantic Victorians: English Literature, 1824–1840*. New York: Palgrave, 2002.

Daly, Nicholas. *Literature, Technology, Modernity, 1860–2000*. Cambridge: Cambridge UP, 2004.

Darby, Margaret Flanders. "Joseph Paxton's Water Lily." In *Bourgeois and Aristocratic Cultural Encounters in Garden Art, 1550–1850*. Edited by Michel Conan, 255–83. Washington, DC: Dumbarton Oaks Research Library and Collection, 2002.

Dawkins, Richard. *The Selfish Gene*, Thirtieth Anniversary Edition. Oxford: Oxford UP, 2006.

de Almeida, Hermione, and George H. Gilpin. *Indian Renaissance: British Romantic Art and the Prospect of India*. Aldershot, UK: Ashgate, 2006.

De Mul, Jos. *Romantic Desire in (Post)Modern Art and Philosophy*. Albany: State University of New York Press, 1999.

De Nooy, Wouter. "Fields and Networks: Correspondence Analysis and Social Network Analysis in the Framework of Field Theory." *Poetics* 31, no. 5 (2003): 305–27.

Deguchi, Yasuo. "Introduction." In *The Examiner – Volume I: 1808*, London: William Pickering, 1996.

Dempsey, Lorcan. "Names and Identities: Looking at Flann O'Brien." Lorcan Dempsey's Weblog: On Libraries, Services and Networks, July 2, 2013. Accessed April 28, 2014. http://orweblog.oclc.org/archives/002212.html.

Derrida, Jacques. *Archive Fever: A Freudian Impression*. Translated by Eric Prenowitz. Chicago: University of Chicago Press, 1998.

DeZelar-Tiedman, Christine. "The Proportion of NUC Pre-56 Titles Represented in the RLIN and OCLC Databases Compared: A Follow-Up to the Beall/Kafadar Study." *College and Research Libraries* 69, no. 5 (2008): 401–6.

Dickens, Charles. *Hard Times*. Edited by David Craig. Harmondsworth, UK: Penguin, 1969.

Diestelkamp, Edward J. "Fairyland in London: The Conservatories of Decimus Burton." *Country Life*, May 19, 1983, 1342–44.

Doughan, David. "Queen (1861–1967)." In *Dictionary of Nineteenth-Century Journalism*. Edited by Laurel Brake and Marysa Demoor, 523–24. Ghent: Academia Press and the British Library, 2009.

Drucker, Johanna. "Humanities Approaches to Graphical Display." *Digital Humanities Quarterly* 5, no. 1 (2011). Accessed February 24, 2014. http://www.digitalhumanities.org/dhq/vol/5/1/000091/000091.html.

During, Simon. *Modern Enchantments: The Cultural Power of Secular Magic*. Cambridge, MA: Harvard UP, 2002.

Eberle-Sinatra, Michael. "'A Natural Piety': Leigh Hunt's *The Religion of the Heart*." *Allen Review* 19 (1998): 18–21.

———. *Leigh Hunt and the London Literary Scene*. Routledge, 2005.

Earhart, Amy E. "Can Information Be Unfettered? Race and the New Digital Humanities Canon." In *Debates in the Digital Humanities*. Edited by Matthew K. Gold. Minneapolis: University of Minnesota Press, 2012:,309–18.

Easley, Alexis. "*Tait's Edinburgh Magazine* in the 1830s: Dialogues on Gender, Class, and Reform." *Victorian Periodicals Review* 38, no. 3 (2005): 263–79.

Ehnes, Caley. "Religion, Readership, and the Periodical Press: The Place of Poetry in *Good Words*." *Victorian Periodicals Review* 45, no. 4 (Winter 2011): 466–87.

Eliot, George. *Middlemarch*. Oxford: Clarendon, 1987.

Ellman, Richard. *The Consciousness of Joyce*. London: Faber and Faber, 1977.

Eyre-Todd, George, ed. *The Autobiography of William Simpson, R. I.* London: T. Fisher Unwin, 1903.

The Examiner, January 11, 1829.

Ezell, Margaret. *Writing Women's Literary History*. Baltimore: John Hopkins UP, 1996.

Fane, Violet. "Hazely Heath." *Woman's World*, November 1887, 10.

Felluga, Dino Franco. "Addressed to the NINES: The Victorian Archive and the Disappearance of the Book." *Victorian Studies* 48, no. 2 (2006): 305–19.

Fish, Stanley. "The Digital Humanities and the Transcending of Mortality." *The New York Times*, January 9, 2012. Accessed April 28, 2014. http://opinionator.blogs.nytimes.com/2012/01/09/ the-digital-humanities-and-the-transcending-of-mortality/.

Fisher, Mark. *Capitalist Realism: Is There No Alternative?* London: Zero Books, 2009.

Fogle, Stephen. *Leigh Hunt's Autobiography: The Earliest Sketches*. Florida: U of Florida P, 1959.

Folsom, Ed. "Database as Genre: The Epic Transformation of Archives." *Publications Of The Modern Language Association Of America* 122 (2007): 1571–79.

Forster-Hahn, Françoise. "Marey, Muybridge and Meissonier: The Study of Movement in Science and Art." *Eadweard Muybridge: The Stanford Years, 1872–1882*. Stanford: Department of Art, Stanford University, 1972. 85–109.

Foucault, Michel. "What Is an Author?" Translated by Donald F. Bouchard and Sherry Simon. In *Language, Counter-Memory, Practice*. Edited by Donald F. Bouchard, 113–138. Ithaca, NY: Cornell UP, 1977.

Fraistat, Neil. "The Function of Digital Humanities Centers at the Present Time." In *Debates in the Digital Humanities*. Edited by Matthew K. Gold. Minneapolis, University of Minnesota Press, 2012.

Fraser, Hilary, Stephanie Green, and Judith Johnston. *Gender and the Victorian Periodical*. Cambridge: Cambridge UP, 2003.

Freud, Sigmund. *Beyond the Pleasure Principle*. Translated by James Strachey. New York: W.W. Norton, 1962.

———. "On Narcissism: An Introduction." Translated by Joan Rivière. *Sigmund Freud: Collected Papers*. New York: Basic Books, 1959. 5 vols. 4:30–59.

Fried, Michael. *Menzel's Realism: Art and Embodiment in Nineteenth-Century Berlin*. New Haven, CT: Yale UP, 2002.

Friedberg, Anne. *The Virtual Window: From Alberti to Microsoft*. Cambridge, MA: MIT Press, 2006.

Friemel, Thomas N. "Why Context Matters." *Why Context Matters: Applications of Social Network Analysis*. Wiesbaden, DE, VS Verlag für Sozialwissenschaften, 2008. doi:10.1007/978-3-531-91184-7.

Fruchterman, Thomas M. J., and Edward M. Reingold. "Graph Drawing by Force-Directed Placement." *Software: Practice and Experience* 21, no. 11 (1991): 1129–64.

Fulford, Tim. "Virtual Topography: Poets, Painters, Publishers and the Reproduction of the Landscape in the Early Nineteenth Century." *Romanticism and Victorianism on the Net* 57–58 (February–May 2010). Accessed February 12, 2015. http://www.erudit.org/revue/ravon/2010/v/n57-58/1006512ar.html.

Fussell, Paul. *The Great War and Modern Memory*. London: Oxford UP, 1975.

Galperin, William. *The Return of the Visible in British Romanticism*. Baltimore: Johns Hopkins UP, 1993.

Garrett, Edmund H., and Edmund Gosse, eds. *Victorian Songs: Lyrics of the Affections and Nature*. Boston: Little, Brown, 1895.

Garvey, Ellen Gruber. *Writing with Scissors: American Scrapbooks from the Civil War to the Harlem Renaissance*. Oxford: Oxford UP, 2012.

Gaye, Tuchman, and Nina Fortin. *Edging Women Out: Victorian Novelists, Publishers, and Social Change*. New Haven, CT: Yale UP, 1989.

Genette, Gérard. *Paratexts: Thresholds of Interpretation*. Translated by Jane E. Lewin. Literature, Culture, Theory 20. Cambridge: Cambridge UP, 1997. Originally published as *Seuils* (Paris: Édition de Seuil, 1987).

"The Getty Thesaurus of Geographic Names Online." The Getty Research Institute. n.d. Accessed May 5, 2014. http://www.getty.edu/research/tools/vocabularies/tgn/index.html.

Gibbs, Frederick W., and Daniel J. Cohen. "A Conversation with Data: Prospecting Victorian Words and Ideas." *Victorian Studies* 54, no. 1 (Autumn 2011): 69–77.

Gibson, Helen, Joe Faith, and Paul Vickers. "A Survey of Two-Dimensional Graph Layout Techniques for Information Visualisation." *Information Visualization* 12, no. 3–4 (2012): 324–57. doi:10.1177/1473871612455749.

Gillington, Alice E. "Nocturnes." In *Women Poets of the Victorian Era*. Edited by Elizabeth A. Sharp, 284–86. London: Walter Scott, 1890.

Gillington, Alice E. "The Doom-Bar." In *A Victorian Anthology 1837–1895: Selections Illustrating the Editor's Critical View of British Poetry in the*

Reign of Victoria. Edited by E. C. Stedman, 609. Boston: Houghton, Mifflin, and Co., 1895.

———. "The Rosy Musk-Mallow." In *A Victorian Anthology 1837–1895: Selections Illustrating the Editor's Critical View of British Poetry in the Reign of Victoria*. Edited by E. C. Stedman, 608. Boston: Houghton, Mifflin, and Co., 1895.

———. "The Seven Whistlers." In *A Victorian Anthology 1837–1895: Selections Illustrating the Editor's Critical View of British Poetry in the Reign of Victoria*. Edited by E. C. Stedman, 608. Boston: Houghton, Mifflin, and Co., 1895.

———. "A West-Country Love-Song." In *Women Poets of the Victorian Era*. Edited by Elizabeth A. Sharp, 283. London: Walter Scott, 1890.

———. "A West-Country Love-Song." In *Women's Voices: An Anthology of the Most Characteristic English, Scotch, and Irish Women*. Edited by Elizabeth A. Sharp, 408. London: Walter Scott, 1887.

Gillington, M. C. "'All Among the Barley.'" *Woman's World*, September 1889, 573.

———. "Amabel at Work," *Woman's World*, August 1889, 521.

———. "Atlantis." In *Women's Voices: An Anthology of the Most Characteristic English, Scotch, and Irish Women*. Edited by Elizabeth A. Sharp, 404. London: Walter Scott, 1887.

———. "Christmas Eve." *Woman's World*, October 1889, 664.

———. "Cloudy Skies." *Woman's World*, April 1889, 310.

———. "A Complaint." *Woman's World*, September 1889, 610.

———. "A Dead March." In *Women Poets of the Victorian Era*. Edited by Elizabeth A. Sharp, 278–79. London: Walter Scott, 1890.

———. "The First Night of the Year." *Woman's World*, January 1889, 155.

———. "A Grey Day." *Woman's World*, October 1889, 632.

———. "Gusty Weather." *Woman's World*, March 1889, 271.

———. "The Home Coming." In *Women Poets of the Victorian Era*. Edited by Elizabeth A. Sharp, 280–82. London: Walter Scott, 1890.

———. "The Home Coming." In *Women's Voices: An Anthology of the Most Characteristic English, Scotch, and Irish Women*. Edited by Elizabeth A. Sharp, 405–7. London: Walter Scott, 1887.

———. *May Byron's How-to-Save Cookery: A War-Time Cookery Book*. London: Hodder and Stoughton, 1915.

———. *May Byron's Rations Book*. London: Hodder and Stoughton, 1918.

———. "A Morning Meeting." *Woman's World*, June 1889, 405.

———. *Pot-Luck, or the British Home Cookery Book*. London: Hodder and Stoughton, 1914.

———. "St. Valentine's Day." *Woman's World*, February 1889, 198.

———. "The Shearing at the Stepping-Stones." *Woman's World*, May 1889, 381.

———. "Summer Night." *Woman's World*, July 1889, 464.

The Girl with the Dragon Tattoo (original title *Män som hatar kvinnor*), film, directed by Niels Arden Oblev. Yellow Bird, Sweden, 2009.

Gitelman, Lisa. *Always Already New: Media, History and the Data of Culture.* Cambridge, MA: MIT Press, 2008.

Gitelman, Lisa, and Virginia Jackson. Introduction to "*Raw Data" Is an Oxymoron.* Edited by Lisa Gitelman, 1–14. Cambridge: MIT Press, 2013.

Gleadle, Kathryn Jane. *The Early Feminists: Radical Unitarians and the Emergence of the Women's Rights Movement, c. 1831–1851.* New York: St. Martin's Press, 1995.

Golbeck, Jennifer, Gilberto Fragoso, Frank Hartel, Jim Hendler, Jim Oberthaler, and Bijan Parsia. "The National Cancer Institute's Thesaurus and Ontology." *Web Semantics: Science, Services and Agents on the World Wide Web* 1, no. 1 (2011).

Gorky, Maxim. "In the Kingdom of Shadows." In *In the Kingdom of Shadows: A Companion to Early Cinema.* Edited by Colin Harding and Simon Popple, 5–6. London: Cygnus Arts, 1996.

Gosse, Edmund. "A Plea for Certain Exotic Forms of Verse." *The Cornhill Magazine*, July 1877, 53.

Green, Stephanie. "Oscar Wilde's *The Woman's World.*" *Victorian Periodicals Review* 3, no. 2 (Summer 1997): 102–20.

Green-Lewis, Jennifer. *Framing the Victorian: Photography and the Culture of Realism.* Ithaca, NY: Cornell UP, 1996.

Griffiths, Alison. *Shivers Down Your Spine: Cinema, Museums, and the Immersive View.* New York: Columbia UP, 2008.

Grossman, Lev, and Gary Moskowitz. "A Handmade World." *Time* 174 (2009): 82–84.

Guillory, John. *Cultural Capital: The Problem of Literary Canon Formation.* Chicago: The University of Chicago Press, 1993.

Gunning, Tom. "'We Are Here and Not Here': Late Nineteenth-Century Stage Magic and the Roots of Cinema in the Appearance (and Disappearance) of the Virtual Image." In *A Companion to Early Cinema.* Edited by André Gaudreault, Nicolas Dulac, Santiago Hidalgo, 52–64. Oxford: John Wiley, 2012.

———. "The Whole Town's Gawking: Early Cinema and the Visual Experience of Modernity." *Yale Journal of Criticism* 7, no. 2 (1994): 189–201.

Hager, Lisa. "Aiming to Misbehave at the Boundary between the Human and the Machine: The Queer Steampunk Ecology of Joss Whedon's *Firefly* and *Serenity.*" In *The Philosophy of Joss Whedon.* Edited by Dean Kowalski and Evan Kreider, 182–93. Lexington, KY: UP of Kentucky, 2011.

Hahn, H. Hazel. "Indian Princes, Dancing Girls and Tigers: The Prince of Wales's Tour of India and Ceylon, 1875–1876." *Postcolonial Studies* 2, no. 2 (2009): 173–92.

Halberstam, J. Jack. *Female Masculinity.* Durham, NC: Duke UP, 1998.

Hansen, Miriam Bratu. *Cinema and Experience: Siegfried Kracauer, Walter Benjamin, and Theodor W. Adorno.* Berkeley: University of California Press, 2012.

Haraway, Donna J. *Simians, Cyborgs, and Women: The Reinvention of Nature.* New York: Routledge, 1991.

———. "Situated Knowledges: The Science Question in Feminism and the Privilege of Partial Perspective." *Feminist Studies* 14, no. 3 (1988): 575–99.

Hardcastle, Ephraim [William Henry Pyne], ed. *Somerset House Gazette, and Literary Museum; or, Weekly Miscellany of Fine Arts, Antiquities, and Literary Chit Chat.* 2 vols. London: W. Wetton, 1824.

Harris, Katherine D. "Big Data, DH, Gender: Silence in the Archives?" *triproftri.* Last modified March 3, 2012. Accessed May 5, 2014. http://triproftri.wordpress.com/2012/03/03/big-data-dh-gender-silence-in-the-archives/.

Hayot, Eric. "What Is Data in Literary Studies?" http://erichayot.org/ephemera/mla-what-is-data-in-literary-studies/.

Hazlitt, William. *Sketches of the Principal Picture-Galleries in England.* London: Taylor and Hessey, 1824.

Heineman, Helen. *Frances Trollope.* Boston: Twayne, 1984.

Henderson, Andrea K. *Romanticism and the Painful Pleasures of Modern Life.* Cambridge: Cambridge UP, 2008.

Henisch, Heinz K., and Bridget A. Henisch. *The Photographic Experience 1839–1914: Image and Attitudes.* University Park: The Pennsylvania State UP, 1994.

Heuser, Ryan, and Long Le-Khac. "A Quantitative Literary History of 2,958 Nineteenth-Century British Novels: The Semantic Cohort Method." *Pamphlet 4.* Stanford, CA: Stanford Literary Lab, 2012.

Hoff, Molly. "The Pseudo-Homeric World of *Mrs. Dalloway.*" *Twentieth-Century Literature* 45, no. 2 (Summer 1999): 186–209.

Hogarth, Paul. *The Artist as Reporter.* London: Gordon Fraser Gallery Limited, 1986.

Holden, Anthony. *The Wit in the Dungeon: Leigh Hunt and his Circle*, London and New York: Little, Brown, and Company, 2005.

Holland, Merlin, and Rupert Hart-David, eds. *The Complete Letters of Oscar Wilde.* New York: Henry Holt and Company, 2000.

Hoog, Marie-Jacques. "Trollope's Choice: Frances Trollope Reads George Sand." In *Woman as Mediatrix: Essays on Nineteenth-Century Women Writers.* Edited by Avriel H. Goldberger and Germaine Brée, 59–72. Westport, CT: Greenwood, 1987.

Houston, Natalie M. "Toward a Computational Analysis of Victorian Poetics." *Victorian Studies* 56, no. 3 (2014): 498–510.

Hu, Yifan. "Efficient, High-Quality Force-Directed Graph Drawing." *Mathematica Journal* 10, no. 1 (2005): 37–71.

Hughes, Linda K., and Michael Lund. *The Victorian Serial.* Charlottesville: UP of Virginia, 1991.

Hughes, Linda K. "What the *Wellesley Index* Left Out: Why Poetry Matters to Periodical Studies." *Victorian Periodicals Review* 40, no. 2 (2007): 91–125.

Hunt, Leigh. *Lord Byron and Some of his Contemporaries.* London: Henry Colburn, 1828.

———. *The Autobiography of Leigh Hunt; with Reminiscences of Friends and Contemporaries*, 3 vols. London: Smith, Elder and Co., 1850.
———. *The Autobiography of Leigh Hunt*. Ed. J. E. Morpurgo. London: The Cresset Press, 1949.
———. *The Selected Writings of Leigh Hunt*, Edited by M. Eberle-Sinatra and R. Morrison, 6 vols. London: Pickering & Chatto, 2003.
Hunter, Fred. "Meteyard, Eliza (1816–1879)." *Oxford Dictionary of National Biography*. Oxford: Oxford UP, 2004. Accessed April 28, 2014. doi:10.1093/ref:odnb/18624.
Hyde, Ralph. *Panoromania! The Art and Entertainment of the "All-Embracing" View*. London: Trefoil, 1988.
———. *The Regent's Park Colosseum*. London: Ackerman, 1982.
Hyde, Ralph, and Valerie Cumming. "The Prints of Benjamin Read, Tailor and Printmaker." *Print Quarterly* 17 (2000): 262–84.
IFLA Study Group on the Functional Requirements of Bibliographic Records. *Functional Requirements of Bibliographic Records: Final Report*. IFLA Series on Bibliographic Control 19. Munich: K. G. Saur Verlag, 1998. http://www.ifla.org/publications/functional-requirements-for-bibliographic-records.
The Illustrated London News, October 14, 1843.
The InPhO Project. n.d. Accessed May 5, 2013. https://inpho.cogs.indiana.edu/.
Jackson, Mason. *The Pictorial Press: Its Origin and Purpose*. London: Hurst & Blackett, 1885.
Jaensch, E. R. *Eidetic Imagery and Typological Methods of Investigation*. Westport, CT: Greenwood Press, 1930, reprinted 1970.
Jauss, Hans Robert. *Towards an Aesthetic of Reception*. Minneapolis: University of Minnesota Press, 1982.
Jenkins, Henry, and Sam Ford. *Spreadable Media: Creating Value and Meaning in a Networked Culture*, Kindle Edition. New York: NYU Press, 2013.
Jeter, K. W. "Letter." *Locus* 20, no. 4 (1987): 57.
Jockers, Matthew L. *Macroanalysis: Digital Methods and Literary History*. Champaign: University of Illinois, 2013.
Johnston, Johanna. *The Life, Manners, and Travels of Fanny Trollope: A Biography*. New York: Hawthorn Books, 1978.
Jones, Owen. *The Grammar of Ornament*. London: Bernard Quaritch, 1868 [1856].
Jordan, Philip D. "Some Lincoln and Civil War Songs." *The Abraham Lincoln Quarterly*, September 1942, 138.
Kennedy, J. Gerald. "Introduction." In *Poe and the Remapping of Antebellum Print Culture*. Edited by Kennedy and Jerome McGann, 1–12. Baton Rouge: Louisiana State UP, 2012.
Kent, Christopher. "The Whittington Club: A Bohemian Experiment in Middle Class Social Reform." *Victorian Studies* 18, no. 1 (1974): 31–55.
Kipling, Rudyard. "Mrs. Bathurst." *The Writings in Prose and Verse of Rudyard Kipling*. New York: Charles Scribner's Sons, 1904. 28 vols. 22:379–408.

Kracauer, Siegfried. *The Theory of Film: The Redemption of Physical Reality.* Oxford: Oxford UP, 1997.

Labbe, Jacqueline M. *Romantic Visualities: Landscape, Gender, and Romanticism.* Basingstoke, UK: Macmillan Press; New York: St. Martin's Press, 1998.

Landré, Louis. *Leigh Hunt (1784–1859) Contributions à l'histoire du romantisme anglais*, 2 vols. Paris: Edition Les Belles-Lettres, 1936.

Latour, Bruno. *Reassembling the Social: An Introduction to Actor-Network-Theory.* Oxford: Oxford UP, 2005.

Ledbetter, Kathryn. "Time and the Poetess: Violet Fane and the Fin-de-Siècle Poetry in Periodicals." *Victorian Poetry* 52, no. 1 (Spring 2014): 141–60.

Levy, Amy. "The Recent Telepathic Occurrence at the British Museum." *Woman's World*, November 1887, 31–32.

Lightbown, Ronald W. "Introduction." *The Life of Josiah Wedgwood: From His Private Correspondences and Family Papers in the Possession of Joseph Mayer, F. Wedgwood, C. Darwin, Miss Wedgwood and Other Original Sources: With an Introductory Sketch of the Art of Pottery in England.* London: Cornmarket Press, 1970.

Liu, Alan. "Where Is Cultural Criticism in Digital Humanities?" In *Debates in Digital Humanities.* Edited by Matthew K. Gold, 490–509. Minneapolis: University of Minnesota Press, 2012.

———. *The Laws of Cool: Knowledge Work and the Culture of Information.* Chicago: University of Chicago Press, 2004.

Locke, John. *An Essay Concerning Human Understanding.* Edited by Alexander Campbell Fraser, 331–62. 2 vols. 1894; rpt. New York: Dover, 1959.

Lusk, Lewis. "A Famous Journalist, Sydney P. Hall, M. V. O." *Art Journal* (1905): 277–281.

Mandell, Laura. "How to Read a Literary Visualisation: Network Effects in the Lake School of Romantic Poetry." *Digital Studies/Le champ numérique* 3, no. 2 (2013). Accessed February 12, 2015. http://www.digitalstudies.org/ojs/index.php/digital_studies/article/view/236/304.

Manovich, Lev. *The Language of New Media.* Cambridge: MIT Press, 2001.

Marey, Étienne-Jules. "Natural History of Organized Bodies." *Annual Report of the Board of Regents the Smithsonian Institution for 1867.* Washington: Government Printing Office, 1868. 277–304.

Marshall, Mary A. "Medicine as a Profession for Woman." *Woman's World*, January 1888, 105–10.

Massumi, Brian. "Envisioning the Virtual." In *The Oxford Handbook of Virtuality.* Edited by Mark Grimshaw, 55–70. Oxford: Oxford UP, 2014.

Mavor, Carol. *Pleasures Taken: Performances of Sexuality and Loss in Victorian Photographs.* Durham, NC: Duke UP, 1995.

McClelland, the Reverend Robert. *Heroes and Gentlemen: An Army Chaplain's Experiences in South Africa.* Paisley, UK: J. and R. Parlane, 1902.

McDonald, Peter D. *British Literary Culture and Publishing Practice, 1880–1914.* Cambridge: Cambridge UP, 1997.

McGann, Jerome J., ed. *The New Oxford Book of Romantic Period Verse.* Oxford: Oxford UP, 1993.

McGann, Jerome. "From Text to Work: Digital Tools and the Emergence of the Social Text." *Romanticism on the Net* 41–42 (2006). Accessed December 26, 2014. http://www.erudit.org/revue/ron/2006/v/n41-42/013153ar.html.

———. "The Rationale of Hypertext." http://www2.iath.virginia.edu/public/jjm2f/rationale.html. 1995.

———. *A New Republic of Letters.* Harvard UP, 2014.

McPherson, Tara. "Why Are the Digital Humanities So White? or Thinking the Histories of Race and Computation." In *Debates in the Digital Humanities.* Edited by Matthew K. Gold, 139–60. Minneapolis: University of Minnesota Press, 2012.

Mermin, Dorothy. *Godiva's Ride: Women of Letters in England, 1830–1880.* Bloomington: Indiana UP, 1993.

Meteyard, Eliza. *Struggles for Fame*, vol. 1. London: T. C. Newby, 1845. Google Books edition. Accessed April 28, 2014.

Meteyard, Eliza. *Dora and Her Papa.* London: George Routledge and Sons, 1869. Google Books edition. Accessed April 28, 2014.

Meteyard, Eliza. "Eliza Meteyard Letter to Leigh Hunt, June 26, 1848." Transcribed by Anne Covell. *Leigh Hunt Letters Collection*, University of Iowa Libraries, Iowa City, Iowa. Accessed April 20, 2014. http://digital.lib.uiowa.edu/cdm/search/collection/leighhunt/searchterm/Meteyard,+Eliza,+1816-1879/field/subjec/mode/exact/conn/and/cosuppress/.

Michel, Jean-Baptiste, Yuan Kui Shen, Aviva Presser Aiden, Adrian Veres, Matthew K. Gray, Joseph P. Pickett, Dale Hoiberg, Dan Clancy, Peter Norvig, Jon Orwant, Steven Pinker, Martin A. Nowak, and Erez Lieberman Aiden. "Quantitative Analysis of Culture Using Millions of Digitized Books." *Science* 331, no. 6014 (2011): 176–82.

Milton, John. *The Poems of John Milton.* Edited by John Carey and Alastair Fowler. London and Harlow: Longmans, 1968.

Moore, Bo. "Inside the Epic Online Space Battle That Cost Gamers $300,000." *Wired*, February 8, 2014. Accessed June 20, 2014. http://www.wired.com/2014/02/eve-online-battle-of-b-r/?cid=co18382344.

Moretti, Franco. *The Bourgeois: Between History and Literature.* Brooklyn, NY: Verso, 2013.

———. *Graphs, Maps, Trees: Abstract Models for Literary History.* Brooklyn, NY: Verso, 2005.

The Morning Chronicle, January 10, 1829.

Morning Journal, January 19, 1829; February 15, 1830.

Mulvey, Laura. "Visual Pleasure and Narrative Cinema." *Screen* 16, no. 3 (1975): 6–18.

Mussell, James. *The Nineteenth-Century Press in the Digital Age.* Houndmills, UK: Palgrave, 2012.

Nahon, Karine, and Jeff Hemsely. *Going Viral.* Malden, MA: Polity Press, 2013.

Neville-Sington, Pamela. "Trollope, Frances (1779–1863)." In *Oxford Dictionary of National Biography.* Edited by H. C. G. Matthew and Brian Harrison. Oxford: OUP, 2004. Online edition, edited by Lawrence Goldman, May 2008. Accessed May 28, 2014. http://www.oxforddnb.com.proxy.its.virginia.edu/view/article/27751.

The Newcastle Courant, January 17, 1829.

Nicholson, Bob. "'You Kick the Bucket; We Do the Rest!': Jokes and the Culture of Reprinting in the Transatlantic Press." *Journal of Victorian Culture* 17, no. 3 (2012): 273–86.

NINES: Nineteenth Century Scholarship Online. Accessed April 28, 2014. http://www.nines.org/.

Nowviskie, Bethany. "A Scholar's Guide to Research, Collaboration, and Publication in NINES." *Romanticism and Victorianism on the Net* 47 (2007). Accessed February 12, 2015. http://www.erudit.org/revue/ravon/2007/v/n47/016707ar.html.

Oberholtzer, Ellis Paxson. *Abraham Lincoln.* Philadelphia: George W. Jacobs and Company, 1904.

O'Connell, Lisa. "Dislocating Literature: The Novel and the Gretna Green Romance, 1770–1850." *Novel: A Forum on Fiction* 35 (2001): 5–23.

Oettermann, Stephan. *The Panorama: History of a Mass Medium.* New York: Zone Books, 1997.

Connelly, Brian. "Creating Colorblind-Friendly Figures." *Brian Connelly,* October 16, 2013. Accessed September 21, 2014. http://bconnelly.net/2013/10/creating-colorblind-friendly-figures/.

Okrent, Arika. "The Listicle as Literary Form." *The University of Chicago Magazine,* January/February 2014. Accessed December 26, 2014. http://mag.uchicago.edu/arts-humanities/listicle-literary-form.

The Open Biological and Biomedical Ontologies. Last modified May 5, 2014. Accessed May 5, 2014. http://www.obofoundry.org/.

Osborne, Peter. D. *Travelling Light: Photography, Travel and Visual Culture.* Manchester: Manchester UP, 2000.

Otto, Peter, and Abigail H. Nedeau-Owen. "Humphry Repton: 'View from the House at Repton.'" In "Innovations in Encompassing Large Scenes," *Romantic Circles Gallery.* Accessed June 10, 2014. http://www.rc.umd.edu/gallery.

Otto, Peter. *Multiplying Worlds.* Oxford: Oxford UP, 2011.

Oxford English Dictionary. s.v. "network." Accessed April 28, 2014. http://www.oed.com.

Page, Lawrence, Sergey Brin, Rajeev Motwani, and Terry Winograd. "The PageRank Citation Ranking: Bringing Order to the Web." *Stanford InfoLab* (1999). Accessed May 5, 2014. http://ilpubs.stanford.edu:8090/422/.

Pal, Pratapaditya, and Vidya Dehiejia. *From Merchants to Emperors: British Artists and India 1757–1930.* Ithaca, NY: Cornell UP, 1986.

Palgrave, Francis Turner, ed. *The Golden Treasury of the Best Songs and Lyrical Poems in the English Language*. Cambridge; London: Macmillan, 1861.

Payne, Robert. "Virality 2.0." *Cultural Studies* 27, no. 4 (2013): 540–60.

Percy, Reuben, John Timbs, and John Limbird, eds. *The Mirror of Literature, Amusement, and Instruction: Containing Original Essays*. London: J. Limbird, 1829.

Perschon, Mike. "Steam Wars." *Neo-Victorian Studies* 3, no. 1 (2010): 127–66.

Pinch, Adela. "Rhyme's End." *Victorian Studies* 53, no. 3 (Spring 2011): 485–94.

Portsmouth, Eveline. "The Position of Women." *Woman's World*, November 1887, 7–10.

Postcolonial Digital Humanities. Last modified April 2014. Accessed May 5, 2014. http://www.dhpoco.org.

Poston, Lawrence. "Thomas Adolphus Trollope: A Victorian Anglo-Florentine." *Bulletin of the John Rylands University Library of Manchester* 49 (1966): 133–64.

Price, Leah. *The Anthology and the Rise of the Novel*. Cambridge: Cambridge UP, 2000.

Ramsay, Stephen. *Reading Machines: Toward an Algorithmic Criticism*. Champaign: University of Illinois, 2011.

Ray, Romita. "The Memsahib's Brush." In *Orientalism Transposed: The Impact of the Colonies on British Culture*. Edited by Julie F. Codell and Dianne Sachko Macleod, 89–116. Aldershot, UK: Ashgate, 1998.

Read, Benjamin. "Summer: The Present Fashions." London: B. Read, c.1830. Accessed May 20, 2014. http://www.motco.com/imageone.asp?Randompic=90004007.

Ridolfo, Jim, and Dànielle Nicole DeVoss. "Composing for Recomposition: Rhetorical Velocity and Delivery." *Kairos* 13, no. 2 (Spring 2009). Accessed December 26, 2014. http://kairos.technorhetoric.net/13.2/topoi/ridolfo_devoss/index.html.

Robson, Catherine. "Girls Underground, Boys Overseas: Representations of Dead Children in Nineteenth-Century British Literature." In *Dickens and the Children of Empire*. Edited by Wendy Jacobson, 116–27. London: Macmillan, 2000.

Robson, Catherine. *Heart Beats: Everyday Life and the Memorized Poem*. Princeton, NJ: Princeton UP, 2012.

Rodin, Auguste. *On Art and Artists*. Translated by Paul Gseil. London: Peter Owen, 1958.

Roe, Nicholas. *Fiery Heart: The First Life of Leigh Hunt*. London: Pimlico, 2005.

Rogers, Helen. *Women and the People: Authority, Authorship and the Radical Tradition in Nineteenth-Century England*. Aldershot, UK: Ashgate, 2000.

Rosenberg, Daniel. "Data before the Fact." In *"Raw Data" Is an Oxymoron*. Edited by Lisa Gitelman, 15–40. Cambridge: MIT Press, 2013.

Russell, W. H. *The Prince of Wales' Tour: A Diary of India*. London: Sampson Low, Marston, Searle & Rivington, 1877.

Ryan, James. *Picturing Empire: Photography and the Visualization of the British Empire*. Chicago: University of Chicago Press, 1997.

———. "Images and Impressions: Printing, Reproduction and Photography." In *The Victorian Vision: Inventing New Britain*. Edited by John M. Mackenzie, 215–39. London: V&A Publications, 2001.

Sachdev, Vibhuti, and Giles Tillotson. *Building Jaipur: The Making of an Indian City*. London: Reaktion Books, 2002.

Sala, George Augustus. "*India and the Prince of Wales*." [The Indian Extra Number of *The Illustrated London News*.] London, 1875.

Saler, Michael T. *As If: Modern Enchantment and the Literary Prehistory of Virtual Reality*. Oxford: Oxford UP, 2012.

Sampson, Gary D. "Unmasking the Colonial Picturesque Samuel Bourne's Photographs of Barrackpore Park." In *Colonialist Photography: Imag[in]ing Race and Place*. Edited by Eleanor M. Hight and Gary D. Sampson, 84–106. London: Routledge, 2002.

Seltzer, Mark. *Bodies and Machines*. New York: Routledge, 1992.

Serano, Julia. *Whipping Girl: A Transsexual Woman on Sexism and the Scapegoating of Femininity*. Emeryville, CA: Seal Press, 2007.

Shapiro, Gisèle. "Autonomy Revisited: The Question of Mediations and Its Methodological Implications." *Paragraph* 35, no. 1 (2012): 30–48. doi:10.3366/para.2012.0040.

Sharp, Elizabeth A., ed. *Sea-Music: An Anthology of Poems and Passages Descriptive of the Sea*. London: Walter Scott, 1887.

Shattock, Joanne, ed. *The Cambridge Bibliography of English Literature: 1800–1900*. Vol. 4. Cambridge: Cambridge UP, 1999.

Silverpen [Eliza Meteyard]. "A Winter's Tears." *The Ladies' Companion and Monthly Magazine* 19. London: Rogerson and Tuxford, 1861: 18–23.

Simpson, John, Susan Brown, Jentery Sayers, Jon Bath, Harvey Quamen, Jon Saklofske, Adèle Barclay, Alex Christie, Mandy Elliott, and the INKE Research Group. "The Emergence of Linked Data in the Humanities." Forthcoming.

Simpson, William. *India "Special"*, exh. cat. London: Burlington Gallery, 1876. [Artist's own annotated copy in British Library.]

———. *Notes and Recollections of My Life* (handwritten memoirs in the collection of the National Library of Scotland), 1889.

Singhal, Amit. "Introducing the Knowledge Graph: Things, Not Strings." n.d. Accessed February 12, 2015. http://googleblog.blogspot.ca/2012/05/introducing-knowledge-graph-things-not.html.

Sloterdijk, Peter. *In the World Interior of Capital*. Translated by Wieland Hoban. Cambridge: Polity Press, 2013.

Smith, David A., Ryan Cordell, and Elizabeth Maddock Dillon. "Infectious Texts: Modeling Text Reuse in Nineteenth-Century Newspapers." In *Proceedings of the Workshop on Big Humanities* (IEEE Computer Society Press 2013).

Smith, Martha Nell. "The Human Touch, Software of the Highest Order: Revisiting Editing as Interpretation." *Textual Cultures* 2, no.1 (2007): 1–15.

Smyth, Brigadier the Rt. Hon. Sir John. *In This Sign Conquer: The Story of the Army Chaplains.* London: Mowbray, 1968.

"The Society of Painters in Water-Colours." *The Art Journal* (London: James S. Virtue, 1861), 173. Google Books edition. Accessed April 28, 2014. https://books.google.com/books?id=7FhVAAAAcAAJ&pg=PA175&dq=%22the+art+journal%22+%22society+of+painters%22&hl=en&sa=X&ei=Wv_dVOupH8L3oATo_4LwBA&ved=0CB8Q6AEwAA#v=onepage&q=%22the%20art%20journal%22%20%22society%20of%20painters%22&f=false.

Starobinski, Jean. *Diderot dans l'espace des peintres ; suivi de, Le sacrifice en rêve.* Paris: Réunion des musées nationaux, 1991.

Stauffer, Andrew. "Digital Scholarly Resources for the Study of Victorian Literature and Culture." *Victorian Literature and Culture* 39 (2011): 293–303.

Stedman, E. C., ed. *A Victorian Anthology 1837–1895: Selections Illustrating the Editor's Critical View of British Poetry in the Reign of Victoria.* Boston: Houghton, Mifflin, and Co., 1895.

———. "Victorian Poets." *The Century Magazine*, January 1873, 357–64.

———. *Victorian Poets.* London: Chatto and Windus, 1875, revised and expanded edition 1887, 1893.

Steedman, Carolyn. *Dust: The Archive and Cultural History.* Manchester: Manchester UP, 2001.

Stone, Marjorie, and Keith Lawson. "'One Hot Electric Breath': EBB's Technology Debate with Tennyson, Systemic Digital Lags in Nineteenth-Century Literary Scholarship, and the EBB Archive." *Victorian Review* 38, no. 2 (2012): 101–25.

Sullivan, Danny. "Google Still World's Most Popular Search Engine by Far, but Share of Unique Searchers Dips Slightly." Searchengineland.com, February 11, 2013. Accessed May 5, 2014. http://searchengineland.com/google-worlds-most-popular-search-engine-148089.

Taddeo, Julie Ann. "Corsets of Steel: Steampunk's Reimagining of Victorian Femininity." In *Steaming into a Victorian Future.* Edited by Julie Ann Taddeo and Cynthia J. Miller, 43–64. Lanham, MD: Scarecrow Press, 2013.

Tavernier, Bertrand, audio commentary. *The Lumière Brothers' First Films.* Paris: Institut Lumière, 1996. DVD.

Thrift, Nigel. "Pass It On: Towards a Political Economy of Propensity." *Emotion, Space and Society* 1 (2008): 83–96.

———. *Non-Representational Theory: Space | Politics | Affect.* London: Routledge, 2008.

Tillotson, Giles. *The Artificial Empire: The Indian Landscapes of William Hodges.* Richmond, UK: Curzon, 2000.

The Times, April 24, 1789; January 10, 1792; August 2, 1827; January 13, 1829.

Todd, Janet. *Feminist Literary History: A Defence*. Cambridge: Polity Press/Basil Blackwell, 1988.

Tomson, Graham R. "Beauty from the Historical Point of View." *Woman's World*, July and August 1889, 454–59, 536–41.

Trollope, Anthony. *An Autobiography*. Ed. David Skilton. London and New York: Penguin, 1996.

Trollope, Frances. *Domestic Manners of the Americans*. London: Gilbert & Rivington, 1832.

———. *Frances Trollope: Her Life and Literary Work from George III to Victoria*. 2 vols. London: R. Bently and Son, 1895.

———. *The Homes and Haunts of the Italian Poets*. London: Chapman & Hall, 1881. Accessed July 25, 2013. http://hdl.handle.net/2027/uc1.$b400926.

Trollope, Thomas Adolphus. *A Decade of Italian Women*. 2 vols. London: Chapman & Hall, 1859. Accessed May 15, 2013. http://archive.org/details/decadeofitalianw00trol.

———. *Filippo Strozzi: A History of the Last Days of the Old Italian Liberty*. London: Chapman & Hall, 1860. Accessed May 16, 2013. http://catalog.hathitrust.org/Record/008640645.

Tukey, John W. *Exploratory Data Analysis*. Reading, MA: Addison-Wesley, 1977.

Turner, Mark. "Time, Periodicals, and Literary Studies." *Victorian Periodicals Review* 39, no. 4 (2006): 309–16.

"UN Women Ad Series Reveals Widespread Sexism." UN Women, October 21, 2013. Accessed April 28, 2014. http://www.unwomen.org/ca/news/stories/2013/10/women-should-ads.

Unsigned. "Christmas Books." *The Times*, December 7, 1888, 13.

Unsigned. "The Woman's World." *The Spectator*, December 7, 1889, 9.

Unsigned. "The Oxford Ladies' Colleges." *Woman's World*, November 1887, 32–35.

Unsigned. "M. C. Gillington." *The Musical Herald*, August 1, 1904, 227–29.

VanderMeer, Ann, and Jeff VanderMeer. "What Is Steampunk?" In *Steampunk II: Steampunk Reloaded*. Edited by Ann VanderMeer and Jeff VanderMeer, 9–16. San Francisco: Worzilla, 2010.

Van Wyk Smith, Malvern. *Drummer Hodge: The Poetry of the Anglo-Boer War 1899–1902*. Oxford: Clarendon Press, 1978.

Wagner, Tamara S. "Frances Trollope [Special Issue]." *Women's Writing* 18, no. 2 (May 2011): 153–292.

Walker, Hugh. *The Greater Victorian Poets*. London: Macmillan 1895.

Watkins, Owen Spencer. *Chaplains at the Front: Incidents in the Life of a Chaplain during the Boer War 1899–1900*. London: S. W. Partridge & Co., 1901.

———. *With French in France and Flanders: Being the Experience of a Chaplain Attached to a Field Ambulance.* London: C. H. Kelly, 1915.

———. *With Kitchener's Army: Being a Chaplain's Experience with the Nile Expedition.* London: S. W. Partridge & Co., 1899.

Webb, Timothy. "Correcting the Irritability of His Temper: The Evolution of Leigh Hunt's *Autobiography*." In *Romantic Revisions.* Edited by Robert Brinkley and Keith Hanley. Cambridge: Cambridge UP, 1992.

Wedgewood, Julia. "Woman and Democracy." *Woman's World*, June 1888, 337–40.

Wilde, Oscar. *The Complete Works.* Vol. 4. Edited by Josephine Guy. Oxford: Oxford UP, 2007.

Wernimont, Jacqueline. "Whence Feminism? Assessing Feminist Interventions in Digital Literary Archives." *Digital Humanities Quarterly* 7, no. 1 (2013). Accessed April 28. http://digitalhumanities.org:8080/dhq/vol/7/1/000156/000156.html.

Westerfeld, Scott. *Behemoth. Leviathan* Series. New York: Simon Pulse, 2010.

Westerfeld, Scott. "Bonus Goliath Chapter and Art!" Scottwesterfeld.com, December 16, 2011. Accessed August 20, 2014, http://scottwesterfeld.com/blog/2011/12/bonus-goliath-chapter-and-art/.

Westerfeld, Scott. *Goliath. Leviathan* Series. New York: Simon Pulse, 2011.

Westerfeld, Scott. *Leviathan. Leviathan* Series. New York: Simon Pulse, 2009.

Westney, Lynn C. "Intrinsic Value and the Permanent Record: The Preservation Conundrum." *OCLC Systems and Services* 23, no. 1 (2007): 5–12.

White, Stephen K. *Edmund Burke: Modernity, Politics and Aesthetics.* Thousand Oaks, CA: Sage Publications, 1994.

Wilcox, Scott. *Edward Lear and the Art of Travel.* New Haven, CT: Yale Center for British Art, 2000.

Wilde, Oscar. "English Poetesses." *Queen*, December 8, 1888, The Poetess Archive. Accessed February 10, 2015. http://idhmc.tamu.edu/poetess/works/wilde1888.html.

———. "Literary and Other Notes." *Woman's World*, November 1887, 36–40.

Wilson, Angus. *The Strange Ride of Rudyard Kipling: His Life and Works.* London: Secker and Warburg, 1977.

Wolfe, Charles. *Remains of the Late Rev. Charles Wolfe.* Edited by the Rev. John A. Russell. Hartford, CT: Huntington, 1828.

Wood, Gillen D'Arcy. *The Shock of the Real: Romanticism and Visual Culture, 1760–1860.* Houndmills, UK: Palgrave, 2001.

Wood, James. "The Slightest Sardine." Review of *The Oxford English Literary History. Vol. XII: 1960–2000: The Last of England?* by Randall Stevenson. *London Review of Books*, May 20, 2004. Accessed May 5, 2014. http://www.lrb.co.uk/v26/n10/james-wood/the-slightest-sardine. *The World*, April 11, 1789.

Yeldham, Charlotte. *Margaret Gillies RWS, Unitarian Painter of Mind and Emotion, 1803–1887*. Lewiston, NY: Edwin Mellen Press, 1997.

Žižek, Slavoj. "Cyberspace, or the Unbearable Closure of Being." In *Endless Night: Cinema and Psychoanalysis, Parallel Histories*. Edited by Janet Bergstrom, 96–125. Berkeley: University of California Press, 1999, 96–125.

Notes on Contributors

Veronica Alfano is a Faculty Fellow at the University of Oregon. The author of several articles and book chapters on Victorian poetry, which focus in particular on gender and genre, she is currently completing a monograph titled "The Lyric in Victorian Memory: Poetic Remembering and Forgetting from Tennyson to Housman."

Alison Booth is Professor of English at the University of Virginia. Her books include *How to Make It as a Woman* (winner of the Barbara Penny Kanner Award) and *Greatness Engendered.* Her database, Collective Biographies of Women, has been supported by ACLS and NEH grants.

Ruth Brimacombe is Associate Curator at the National Portrait Gallery in London. She received her doctorate from the University of Melbourne in 2008 and is developing her thesis, "Imperial Avatars: Art, India and the Prince of Wales in 1875–6," for publication.

Susan Brown is Professor of English and Theatre Studies at the University of Guelph and Visiting Professor at the University of Alberta. She directs and co-edits the Orlando Project and leads the Canadian Writing Research Collaboratory.

Alison Chapman is Associate Professor of English at the University of Victoria. She recently published *A Rossetti Family Chronology* (Palgrave 2007); she is the Director of the Victorian Poetry Network and the Editor of the Database of Victorian Periodical Poetry.

Ryan Cordell is Assistant Professor of English at Northeastern University and Core Founding Faculty Member in the NULab for Texts, Maps, and Networks (http://nulab.neu.edu). His scholarship focuses on convergences among literary, periodical, and religious culture in antebellum American mass media.

Lisa Hager is Assistant Professor of English and Women's Studies at the University of Wisconsin–Waukesha. Her current book project looks at the relationship between "the New Woman" and the Victorian family, and she has published articles on Victorian sexology, the New Woman, aesthetics, steampunk, queer studies, and digital humanities. Lisa is the founding editor for the *Journal of Victorian Culture Online.*

Natalie M. Houston is Associate Professor of English at the University of Houston. She is the Project Director for the Visual Page, an NEH-funded project to develop a software application to identify and analyze visual features in digitized printed books. She is currently writing a monograph entitled "Reading Victorian Poetry Digitally."

Christopher Keep is Associate Professor in the Department of English at the University of Western Ontario. His articles on nineteenth-century literature and culture have appeared in numerous journals and in several collections of essays.

Peter Otto is ARC Research Professor (DORA) at the University of Melbourne. His recent publications include *Multiplying Worlds: Romanticism, Modernity, and the Emergence of Virtual Reality* (OUP 2011).

Catherine Robson is Professor of English at New York University, where she specializes in nineteenth-century British culture and literature. Her *Heart Beats: Everyday Life and the Memorized Poem* just won the North American Victorian Studies Association's Best Book of the Year award.

Michael E. Sinatra is Associate Professor of nineteenth-century studies at the Université de Montréal. He is the author of *Leigh Hunt and the London Literary Scene, 1805–1828* and the founding editor of *Romanticism and Victorianism on the Net.*

Andrew Stauffer is Associate Professor of English at the University of Virginia, where he also directs the digital federation NINES and runs the Book Traces project. He is the author of *Anger, Revolution, and Romanticism* and numerous articles on nineteenth-century literature.

Index

Printed by Printforce, the Netherlands